Advance Praise for

Alcohol Fuel

We should salute the pioneers who worked out the details of small-scale fuel production. We should also do them the honor of turning their innovation into a meaningful agent of change in the future. This book improves our opportunity to do both of those things.

— Bryan Welch, from the Foreword

Finally, a definitive guide to ethanol for sustainable communities and do-it-yourselfers! *Alcohol Fuel* is a splendid brew of theory, practice, and hands-on fundamentals from an expert author who has truly been there and done that.

Dan Fink — co-author, *Homebrew Wind Power;* contributing author, *Home Power Magazine*, *Back Home Magazine*, and more.

Alcohol Fuel is a practical, well-written book that will give you all the information you need to reduce your dependence on big oil. The fact that the focus of this book is on small or community scale operations makes it even more useful to the average person or group considering ethanol as an alternative fuel. I highly recommend it.

— Greg Pahl, Author of *The Citizen-Powered Energy Handbook: Community Solutions to a Global Crisis,* www.gregpahl.com

Richard Freudenberger's *Alcohol Fuel* is a remarkable exploration of small scale energy production. Part history, part memories, it is an amazing look at how we might sustainably fuel ourselves.

— Lyle Estill, author of *Biodiesel Power* and *Small is Possible;* cofounder of Piedmont Biofuels

Alcohol Fuel is what we've been waiting for — fuel production on the appropriate scale.

— Dr. Jack Martin, CSIT, Appropriate Technology Program, Appalachian State University

Writing about technical subjects that honors both the material and the reader is rare, and when brought forth, seems to be borne from a deep intimacy with the subject and an affection for and belief in the capacity of people. Alcohol Fuel is such a work of deep generosity, and I am thankful to have it.

— Forest Gregg, author of *SVO: Powering Your Vehicle with Straight Vegetable Oil,*

A Guide to Small-Scale Ethanol

Alcohol Fuel

A Guide to Small-Scale Ethanol

Alcohol Fuel

MAKING AND USING ETHANOL AS A RENEWABLE FUEL

RICHARD FREUDENBERGER

NEW SOCIETY PUBLISHERS

Cataloging in Publication Data:
A catalog record for this publication is available from the National Library of Canada.

Cover design by Diane McIntosh.
Sugar beets: iStock/Jason Lugo; Still: Richard Freudenberger; Car: istock/Galina Barskaya

Printed in Canada by Friesens.
First printing September 2009.

Paperback ISBN: 978-0-86571-626-1

Inquiries regarding requests to reprint all or part of *Alcohol Fuel* should be addressed to New Society Publishers at the address below.

To order directly from the publishers, please call toll-free (North America)
1-800-567-6772, or order online at www.newsociety.com

Any other inquiries can be directed by mail to:

New Society Publishers
P.O. Box 189, Gabriola Island, BC V0R 1X0, Canada
(250) 247-9737

New Society Publishers' mission is to publish books that contribute in fundamental ways to building an ecologically sustainable and just society, and to do so with the least possible impact on the environment, in a manner that models this vision. We are committed to doing this not just through education, but through action. This book is one step toward ending global deforestation and climate change. It is printed on Forest Stewardship Council-certified acid-free paper that is **100% post-consumer recycled** (100% old growth forest-free), processed chlorine free, and printed with vegetable-based, low-VOC inks, with covers produced using FSC-certified stock. Additionally, New Society purchases carbon offsets based on an annual audit, operating with a carbon-neutral footprint. For further information, or to browse our full list of books and purchase securely, visit our website at: **www.newsociety.com**

To my colleague, coworker and friend, Clarence Goosen,
who threw heart and soul into ethanol fuel back in the day —
— and without whom this book would not have been written.

Books for Wiser Living recommended by *Mother Earth News*

Today, more than ever before, our society is seeking ways to live more conscientiously. To help bring you the very best inspiration and information about greener, more sustainable lifestyles, *Mother Earth News* is recommending select books from New Society Publishers. For more than 30 years, *Mother Earth News* has been North America's "Original Guide to Living Wisely," creating books and magazines for people with a passion for self-reliance and a desire to live in harmony with nature. Across the countryside and in our cities, New Society Publishers and *Mother Earth News* are leading the way to a wiser, more sustainable world.

Contents

Acknowledgments ... XI
Foreword: by Bryan Welch, Odgen Publications, Inc. /*Mother Earth News* ... XIII
Introduction: Why Alcohol? ... 1
Chapter One: About Alcohol Fuel ... 3
Chapter Two: What You'll Need to Start ... 21
Chapter Three: Federal and State Requirements ... 27
Chapter Four: Do-It-Yourself Economics ... 39
Chapter Five: Feedstocks and Raw Materials ... 45
Chapter Six: Starch Conversion, Sugars and Fermentation ... 63
Chapter Seven: The Distillation Process ... 101
Chapter Eight: Preparation, Fermentation and Distillation Equipment ... 119
Chapter Nine: Alcohol as an Engine Fuel ... 155
Chapter Ten: Case Studies ... 205
Conclusion: What Next? ... 215
Appendix A: True Percentage of Proof Spirit ... 217
Appendix B: Resources and Suppliers ... 221
Glossary of Terms ... 225
Bibliography ... 241
Endnotes ... 245
Index ... 247
About the Author ... 257

Acknowledgments

A number of people helped to make this book possible, through the long process of research, recall and development. I would like to recognize and thank:

Dennis Burkholder, Emerson Smyers, B.V. Alvarez, Ned Doyle, John Vogel, Norman Holland and Beach Barrett, my former colleagues and coworkers at the original *Mother Earth News* magazine who truly broke new ground in grass-roots ethanol research and helped disseminate practical, productive and hands-on information. Some are no longer with us, but the memories remain strong.

Clarence Goosen in particular for being so generous with his research, illustrations and artwork, and for keeping us on the straight and narrow path in bringing small-scale alcohol fuel to fruition.

Don Osby at *BackHome* magazine for helping with his photography and Adobe software skills, and everyone else at *BackHome* for doing without me through the times I was focused on the book.

The New Society crew who put up with my delays and still maintained a civil and professional demeanor: Chris Plant, Ingrid Witvoet, Sue Custance, EJ Hurst, and especially my editor, Linda Glass, who has a rare gift of comprehending the English language and the vernacular of arcane technology.

Bill Kovarik, Professor of Communication at Radford University in Virginia, who gave me his valuable time, and an incredible insight into the history of ethanol motor fuels in the 20th century.

Craig Williams, Agriculture Educator of Tioga County, Pennsylvania who introduced me to John Painter of Painterland Farms and Peggy Korth, the Sustainable Agriculture Research and Education coordinator, to complete Chapter 10.

Finally, to Angela Fernandini for keeping the dogs from underfoot and the cats out of my hair while I worked through the final days of the manuscript.

Foreword

Alcohol fuels ignite in the human imagination as readily as they explode in the cylinders of an internal-combustion engine. And they power the human mind with the same eloquent volatility that pushes a 7,000-lb pickup down the freeway. Just strike up a conversation with anyone who's brewing their own fuel and you'll see.

It is powerful stuff.

Richard Freudenberger and I have witnessed the power of alcohol fuels firsthand. During our parallel careers publishing magazines and books about sustainability, we've seen race cars and tractors powered by fuels made from sugar cane and corn. More importantly, we've seen the light in the eyes of people who have discovered the deceptive simplicity and extraordinary power of alcohol fuels at work.

In the 1970s our generation was looking for ways of living more conscientiously. We deplored the environmental and social damage being done by the international petroleum industry in those days and we were eager for alternatives. Searching the landscape for a better source of fuel, we found some resourceful pioneers who had figured out ways of turning surplus agricultural products into affordable, locally produced engine fuel. Eureka! We were energized by the idea, and a few of us made it the focus of our lives and careers.

The energy that created this book came from many different sources, no doubt, but the initiative that led to its creation was powered by the human imagination.

It's the fundamental miracle of photosynthesis that turns us on, I think. The very notion that a plant can take the sun's energy, generated in a fiery nuclear furnace almost 100 million miles away, and store it as stems, leaves and sugars is an almost magical concept. To see those plants converted to a clean, volatile liquid fuel that can propel an automobile only enhances the magic.

It's important not to be distracted by the recent involvement of multinational energy companies and agribusinesses in the creation of ethanol. Those businesses have served a purpose in the development of alcohol as fuel source, but the most profound value in alcohol-based fuel is in its remarkable flexibility outside the worldwide energy industry. As this book beautifully illustrates, alcohol fuels can be created anywhere there is a surplus of energetic plant material, by anyone with a few simple tools and some common sense. In the local economy many common crops can provide the energy needed to fuel local business. If we use our imaginations it's not too hard to envision a world in which many communities provide their own energy for transportation and agriculture, creating a kind of self-sufficiency and security rarely seen in the world today. Many countries where fuel is prohibitively expensive could enhance their national economies, cleanse their air and water and improve the quality of life for their citizens by adopting a few simple technologies for turning surplus crops into alcohol fuels.

Any farmer who's smelled surplus corn rotting in the elevators knows that it stinks like energy. Any chemistry student who's ever fired up a burner with the vapor off a beaker of lab-brewed alcohol has experienced that thrill of seeing the sun's energy reignited before our very eyes.

The more subtle qualities of bio-based fuels are in their ability to replace petroleum products with cleaner, more sustainable substitutes. Increasingly, we are becoming aware of the true cost of getting coal, oil and gas out of the ground and into the economy. We've also begun to contemplate the social and economic costs we will realize as fossil fuels become more and more scarce. Our dependence on foreign petroleum puts us in a precarious position. Fossil fuels are not a sustainable energy source. Biological fuels like alcohol give us invaluable resources to help meet the challenge of humanity's next few decades on the planet, and they give us a chance to change the way we look at the energy industry — maybe even to re-imagine it powered by locally grown, locally brewed fuels that don't have to be transported halfway across the globe.

We don't know which sources of energy may fuel human enterprise tomorrow. But we know the primary sources of our energy today can't last. Alcohol offers a number of advantages that guarantee its importance in the transition we're about to begin. Those qualities also suggest that small-scale alcohol production may be an efficient and responsible source of energy for the long-term future.

We should salute the pioneers who worked out the details of small-scale fuel production. We should also do them the honor of turning their innovation into a meaningful agent of change in the future. This book improves our opportunity to do both of those things.

Bryan Welch
Publisher & Editorial Director
Ogden Publications, Inc.
Mother Earth News, Utne Reader,
Natural Home, The Herb Companion,
Farm Collector, Gas Engine Magazine ,
Grit Magazine, Capper's Magazine

Introduction: Why Alcohol?

The fact that you're reading this reveals that you have at least a passing interest in alcohol as a fuel. For some, that interest may be one of curiosity — about alternatives to petroleum in general or about the practicability of fuel alcohol in particular. Others may be considering using alcohol fuel — better known as ethanol — to supplement or even replace petroleum in their vehicles, tractors, generators, or to power equipment at home or on the farm.

The majority of people who read this book from start to finish will likely learn a lot more about alcohol than they ever suspected there was to know. Alcohol has a long and diverse history, certainly as a beverage, but equally so as a fuel, which only came into sharp focus toward the middle of the 19th century when it was used for lighting. Alcohol's journey from an illuminant to a motor fuel to a clean-air gasoline additive and most recently to a bridge technology slated to help us transition to a petroleum-free age is an absorbing one. The tale is fraught with political chicanery, the impact of wars, industrial espionage, and the sheer energy of a grass-roots movement. Yet the more important story is the fact that an entire litany of common carbohydrates — not just food crops, but agricultural cull, food industry waste, and plants normally considered a nuisance — can actually be turned into a viable fuel, effectively free from the constraints of the marketplace, suppliers, middlemen and tariffs if you so choose.

Depending upon how you approach alcohol fuel production, it is entirely possible to maintain a completely self-sufficient, self-sustaining, environmentally responsible operation able to produce not only fuel, but also valuable co-products that can be sold, bartered or recycled back into your own venture. In this way, alcohol has some real advantages over other renewable fuels in that it doesn't need much, if any, "mainstream" input unless you decide to include it. Having freedom from the unpredictable swings of traditional commodities markets can be a real benefit to long-term planning and peace of mind.

I have intentionally approached the subject of fuel alcohol with small-scale production in mind. Chapters 3, 4 and 8 address this in more detail, but I will summarize it by saying that in steering clear of the conventional industrial-level mindset, a fuel producer can shed him or herself of what's become a stigma associated with relying on corn as a feedstock — that of using food for fuel. You'll see more on that later, but the point is, appropriate-scale production means opening your mind to the vast possibilities that exist, embracing experimentation, and adjusting your approach to manufacturing accordingly. When I first became involved with fuel alcohol at *The Mother Earth News* in 1978, the price of corn and its impact on the global market was nowhere near the issue that it is today. Yet we still didn't simply accept corn and grains as the status quo — potatoes, apple waste, and other atypical feedstocks were fair game for the sheer flexibility of supply.

Now, the question often arises at seminars and presentations as to how small is small? That, again, is up to you. In the chapter on distillation equipment, I discuss a plain-vanilla batch still capable of making a few gallons of fuel-grade ethanol per hour. Something like that will barely serve the needs of a typical motorist, but there are some good arguments for considering such a piece of equipment in the experience it can provide, devoid of a hefty investment. People in exurban or rural areas can easily contemplate a larger still, fully capable of output in the range of 12 to 35 gallons per hour or more, without breaking the bank with an automated, continuous-distillation design. And if you consider the options opened with a cooperative venture — be it in an urban or rural environment — you can achieve an economy of scale that makes a larger investment in equipment much more appealing.

I have tried diligently throughout the book to avoid overselling the concept of fuel alcohol as a matter of principle. I have been through that before, and it only taints the waters. As I represent my publishing business or address workshop attendees at various renewable energy and sustainability events across the country, it has become clear to me that there is opportunity in the air and, regrettably, no shortage of opportunists ready to make a quick buck. Particularly in the past few years, interest in fuel alcohol has flourished and along with it the carneys, wrapped in the flag and preaching a grass-roots message from the convenient soap-box of sustainability. This may sell books, products and videos but it does little to legitimize the practical production of alcohol fuel.

Make no mistake: alcohol fuel requires an investment in time and equipment and a commitment to study and effort. The rewards can be great, but the promises of a universal magic potion cure-all can also be very beguiling. What is unfortunate is that eager newcomers to renewable energy and a sustainable lifestyle are the ones who will get taken; often enough, they can move forward and carry on unscathed. Yet there will always be those who depart, embittered, blaming the message and not the false messenger. It is my hope that this book will honestly guide those willing to work toward some degree of self-sufficiency and sustainability in the right direction.

ONE
About Alcohol Fuel

Its history as an energy source • The farmer's "Fuel of the Future" • A promise of renewable energy • Ethanol and the environment • Gasohol and E-85 • Brazil's booming ethanol program • The economics of ethanol • The question of Food vs. Fuel

In his State of the Union speech of January 23, 2007 President George W. Bush proposed to reduce gasoline use in the US by 20 percent over the next ten years: "When we do that, we will have cut our total imports by the equivalent of three-quarters of all the oil we now import from the Middle East."[1] The goal was an ambitious one, requiring a significant increase in the supply of alternative fuels to 35 billion gallons by the year 2017 — five times the existing target at the time. A key item of record was ethanol, of which Bush said "We must continue investing in new methods of producing ethanol using everything from wood chips to grasses, to agricultural wastes."

He was hardly the first to say so, and certainly not the first politician to address the issue. Earlier efforts to incorporate ethyl alcohol with gasoline in a 10 or 15 percent blend stemmed more from the farm lobby in Depression-ravaged corn states such as Nebraska, Illinois, and South Dakota than from any endeavor to replace petroleum. Corn prices had dropped to below 25 percent of their pre-Depression levels and the agricultural community was searching for new markets for their products. Subsequent gasohol programs in the late 1970s touted the 10/90 ethanol/gasoline blend as a fuel extender during the second OPEC crisis — though when the price of petroleum eventually dropped, it was promoted as a lead-free octane booster for a cleaner environment.[2]

In the late 1980s, as dire reports about the state of the environment and early alarms over global warming became more prevalent, efforts to advance ethanol as an eventual replacement for gasoline began anew. Aside from environmental factors, our dependence on foreign petroleum sources, creeping fuel prices, and a flagging farm economy combined to make ethyl alcohol the darling of the corn states once again. Senator Tom Daschle of South Dakota teamed up with Kansas senator Bob Dole in 1990 to amend the Clean Air Act with reformulated gasoline provisions that required petroleum refiners to make oxygenated fuels available to US cities with high carbon monoxide levels.[3] The phase-out of tetraethyl lead from gasoline beginning in 1975 adversely affected the combustion of fuel in vehicles, and the amendment was intended to make gasoline burn more completely by introducing additives that contained oxygen. Ethanol fit the bill perfectly.

Yet, as cars and light trucks saw increases in fuel and combustion efficiency with the advent of microprocessor-controlled fuel-injected engines, the need for "oxygenated" fuel became less critical. That need was mitigated in 2005 by a new piece of legislation, the National Energy Policy Act, which mandated the production of 7.5 billion gallons of ethanol to be blended into gasoline by 2012, along with a 51-cent-per-gallon tax credit for the refiners.

Ethanol Fuel's Extraordinary Past

As I write these pages, there is a battle raging between the proponents of ethanol, which include corn growers, their lobbyists, fuel producers and blenders, and equipment manufacturers, and its detractors, a dissimilar mix of environmentalists, petroleum advocates, the food industry, and adherents of global food policy. And, as it turns out, the conflict is far from original.

In 1978, while doing research for "Mother" — my old employer, *The Mother Earth News* — I found out that there was a lot more to alcohol as a fuel source than I'd ever imagined. My knowledge of motor alcohol had until then been limited to the fact that it could be used in early 4-cylinder Fords as an alternative fuel (along with kerosene and other petroleum derivatives) and that General Rommel, Hitler's legendary Desert Fox, had occasionally fueled his trucks and armored vehicles with the cooling fuel in the heat of North Africa's desert during the campaigns of World War II. The former I'd learned from rebuilding a 1931 Ford sedan, the latter from veteran GI's who said they could identify the Germans' proximity by the distinctive odor of alcohol and sauerkraut.

The magazine had jumped with both feet into the home-grown ethanol movement after an editor was sent to interview Lance Crombie, a Minnesota farmer who'd claimed to have been distilling fuel alcohol from corn he'd otherwise take a loss on in the market. The fact that it cost him $2.00 per bushel to grow corn that he could only get $1.60 for at the time — along with uncommonly high fuel prices and a determined curiosity about solar distillation — had prompted the Minnesotan (who happened to hold a doctorate in microbiology and a degree in biochemistry) to dabble in the alternative fuel, for which he was rewarded with the confiscation of his equipment for making alcohol without a permit.

The federal authorities eventually relented, after deciding that the solar distiller was too primitive to qualify as an alcohol still. Nonetheless,

BACKHOME PHOTO COLLECTION

Fig. 1.1: *Early solar distillation research at The Mother Earth News, using a flat plate design developed by farmer Lance Crombie of Minnesota.*

Mr. Crombie was keen on spreading the word that there *was* an alternative to gasoline and fuel oil, and that the answer for many, especially in rural communities, was to grow their own. And we at *Mother Earth*, in the midst of a national fuel crisis, were more than happy to accommodate him.

Our goal was to find out everything there was to know about alcohol fuel and then develop a program that would put the information in low-tech form out to the buying public. A few years before, the magazine had set up a research facility to test and develop ideas related to low-cost transportation and appropriate technology in the solar and renewable energy fields. After an assessment period, the projects were published, and with the price of fuel at an all-time high, publishing articles that would provide practical answers rather than provoke anger had some very real appeal.

The research department was pretty much given free rein to come up with anything that might work. We established early on that flat-plate solar distillation was not effective on anything other than an experimental scale, and eventually moved on to adapting more traditional distillation techniques to our model, soliciting help from academics and picking the brains of everyone from midwest farmers to the distillers at George Dickel, the Tennessee whiskey manufacturer. Our core research team included Clarence Goosen, an Oklahoma native with solid agricultural roots and a formal education in graphic arts and photography; Dennis Burkholder, a Minnesota farm boy with a prolific technical imagination, creative fabricating skills, and a few design patents to his name;

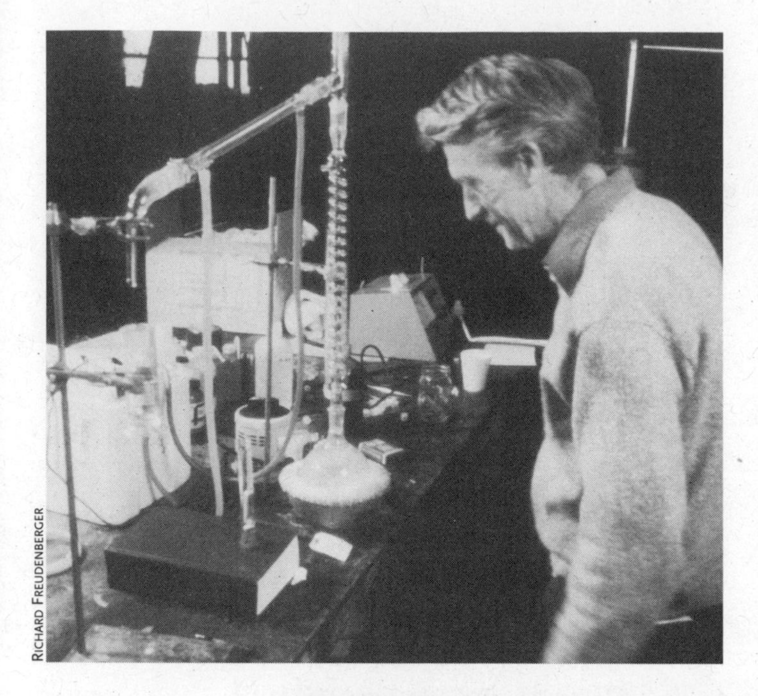

RICHARD FREUDENBERGER

Fig. 1.2: *Initial fuel ethanol experiments at a Mother Earth facility, aided by a volunteer bio-chemist with an interest in farm-based alcohol fuel.*

Emerson Smyers, an old-school machinist and genuine industrial-skills renaissance man; B.V. Alvarez, a former NASCAR-circuit mechanic with an entrepreneurial spirit; and me — charged with organizing the program, putting our discoveries into print, and keeping us on the straight and narrow path with the federal and state authorities.

In those early days there was no Internet, and productive investigation required that you knew your way around academic research and had the ability to do old-fashioned gumshoe footwork. Fortunately, the magazine readers provided a ready-made network of advocates and do-it-yourselfers, although they arrived on the scene with an equal number of swindlers, conspiracy theorists, and outright lunatics. Eventually, with the help of grass-roots enthusiasts, farm

BACKHOME PHOTO COLLECTION

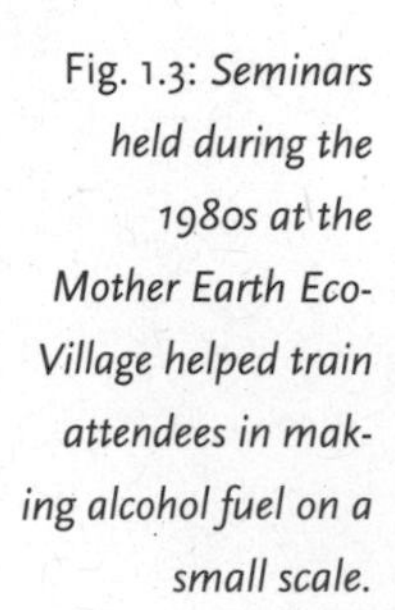

Fig. 1.3: *Seminars held during the 1980s at the Mother Earth Eco-Village helped train attendees in making alcohol fuel on a small scale.*

BackHome photo collection

Fig. 1.4: *Attendees at an alcohol fuel workshop in Michigan, 1979.*

groups, and a proactive Department of Energy, a real alcohol-fuel movement developed, which only dissipated with the policies of the incoming Reagan administration and a change in direction toward industrial-scale gasohol blends.

Mother and her crew of research staff ultimately developed a grass-roots seminar program that we promoted through our Eco-Village (a 624-acre research and educational facility) and through an agenda of traveling seminars, which we took on the road to communities from North Carolina to Nebraska, and Washington, DC to Wisconsin.

You'll notice that some of the photos in this book record the interest and enthusiasm with which our workshops and seminars were received — and I would hope that we haven't become so institutionalized to the national energy system and its inherent price structures

BackHome photo collection

Fig. 1.5: *A promotional alcohol fuel tour to the state capitol at Madison, Wisconsin, 1979.*

RICHARD OWEN

Fig. 1.6: *Mother Earth News' first alcohol-fueled Chevrolet pickup and a NYC taxicab converted to use alcohol fuel set up for a press conference in Washington, DC in 1979.*

that we can't take the initiative to break free from it and develop some individual or local cooperative alternatives once again.

In assembling my research for this book, I was extremely fortunate to happen upon Bill Kovarik, Professor of Communication at Virginia's Radford University and co-author of an early work on ethanol, *The Forbidden Fuel.* Dr. Kovarik enthusiastically shared a good deal of history with me, including insight into early research on ethanol fuel (which was being groomed in the post-World War I era as the replacement to gasoline, believed by some economists, industrialists and geologists alike to be in finite supply), the development of tetraethyl lead as an anti-knock compound despite evidence of its public health hazard, and the Farm Chemurgy movement of the mid-1930s, a populist Republican crusade to develop industrial uses for farm crops, including corn ethanol.

Kovarik's work covers a broad swath of renewable-fuel history and related public policy, and his papers, including "Henry Ford, Charles Kettering and the Fuel of the Future" and "Ethyl: The 1920s Environmental Conflict over Leaded Gasoline" are goldmines of obscure details and particulars of both the populist and industrial attitudes toward the "other" fuel, in the US and abroad.

As early as 1899, the German government was so concerned over its nation's lack of domestic oil reserves and the growing surplus of farm crops that it established the "Centrale fur Spiritus Verwerthung," or Office of Alcohol Sales, to maintain alcohol prices at par with petroleum through subsidies to producers and by placing a tariff on imported oil.[4] In a 1903 survey, 87 percent of German farmers considered alcohol engines to be equal or superior to steam engines in performance. Subsequently, in the period prior to World War I, it's estimated that tens of thousands of distilleries manufactured some 66 million gallons of alcohol per year to benefit the German economy, many of which were small, decentralized farm operations.[5]

Cars like the Model T Ford, manufactured between 1908 and 1927, came equipped with driver-controlled adjustable carburetors and a manual spark advance that made it convenient to switch between gasoline, alcohol and kerosene to suit whatever fuel was available. In 1925, Henry Ford told a New York Times reporter that ethyl alcohol was the fuel of the future, declaring that "[it] is going to come from fruit like that sumach out by the road, or from apples, weeds, sawdust — almost anything." He went on to say that "There is fuel in every bit of vegetable matter that can be fermented. There's enough alcohol in one year's yield of an acre of potatoes to drive the machinery necessary to cultivate the fields for a hundred years."[6]]

RICHARD FREUDENBERGER

Fig. 1.7: *The Model T Ford, manufactured through the 1920s, had a knob on the dashboard to control the setting of the carburetor's main jet.*

RICHARD FREUDENBERGER

Fig. 1.8: *The dashboard control knob is connected through a linkage to the carburetor so the engine can be adjusted to operate on fuel-grade ethanol.*

As internal combustion engines grew to be more sophisticated, it became apparent that the properties of conventional gasoline precluded its use in high-compression designs without some type of anti-knock additive. The use of high-compression engine technology was proving to be the most economical path to engine efficiency and performance, and automotive engineers had reached an impasse of sorts until the fuel question could be resolved. At this point, the attention on alcohol shifted from its use as a petroleum replacement to its value as a clean anti-knock agent that could be blended compatibly with gasoline.

The octane-boosting properties of ethanol are explained in Chapter 9, but it was known even in the early 1920s that alcohol — for reasons mistakenly attributed to its high ignition temperature — could withstand very high compression conditions without knocking or pinging. Testing on alcohol-gasoline and alcohol-benzene mixtures (benzene is an aromatic hydrocarbon derived from coal tar and coke ovens) at various blend ratios was conducted through the early part of the 20th century in the US and in the UK and Europe.

The British Fuel Research Board reported in 1920 that higher engine compression compensated for alcohol's lower caloric value. A mixture of alcohol with 20 percent benzene or gasoline "runs very smoothly, and without knocking." [7] In 1925, a presentation at the New York Chemists Club concluded that "Composite fuels made simply by blending anhydrous alcohol with gasoline have been given most comprehensive service tests extending over a period of eight years. Hundreds of thousands of miles have been covered in standard motor car, tractor, motor boat and aeroplane engines with highly satisfactory results. Alcohol blends easily excel gasoline on every point important to the motorist. The superiority of alcohol-gasoline fuels is now safely established by actual experience." [8] It was during this period that empirical testing procedures were developed to establish reliable references for knock resistance, eventually resulting in the octane rating system we're all now familiar with.

Despite evidence of these early successes, the petroleum industry managed to oppose alcohol blends with considerable vigor. Kovarik's work details this much more thoroughly than I will here, but the upshot is that a business alliance between Standard Oil (now Exxon), General Motors and DuPont Corporation gave birth to the Ethyl Corporation, the sole manufacturer of tetraethyl lead (TEL), a highly toxic compound developed in 1921 and introduced 14 months later as a gasoline additive to prevent engine knock. The use of TEL was licensed under contract to fuel wholesalers, who blended it with gasoline prior to retail delivery. (Whether the firm's founders could have chosen a different brand name to avoid confusion among general public is a moot point; the compound was notated $Pb(C_2H_5)_4$ and legitimately labeled tetraethyl lead. The similarity in name to ethyl alcohol, or ethanol, was unfortunate but a fact of life.)

By the mid-1930s, somewhere between 70 and 90 percent of all gasoline sold contained tetraethyl lead, despite public health concerns and well-documented instances of lead poisoning. Between 1923 and 1925, manufacturing operations in New Jersey killed 15 refinery workers and affected several hundred more by rendering them violently insane. Headlines trumpeted the effects of "looney gas," and for a period, New York City, Philadelphia and municipalities in New Jersey halted the sale of leaded fuel.

In 1925, the US Surgeon General temporarily suspended the production and sale of leaded gasoline, and a panel was appointed to investigate the fatalities and weigh the dangers of widespread distribution of the lead compound through its sale as a gasoline additive. The panel was given seven months to conduct and analyze its tests. The June 1926 final report noted the time constraints given the panel and pointed out the long gestation period required to produce detectable symptoms of lead poisoning, but nonetheless ruled that there were "no good grounds for prohibiting the use of ethyl gasoline ... as a motor fuel, provided that its distribution and use are controlled by proper regulations."[9]

Despite concerns brought up in the report that extensive use of leaded gasoline might alter circumstances and render its use more of a hazard, particularly with regard to less obvious chronic degenerative diseases (and in spite of committee recommendations that the investigation continue, for public safety reasons, in light of the increase in the number of automobiles throughout the country), the Surgeon General's office in 1927 set a voluntary standard of 3 cubic centimeters per gallon for the oil industry to follow in mixing tetraethyl lead with gasoline — a level which corresponded to the maximum then in use among refiners. At the same time, the industry internally took steps toward instituting safer working conditions within the oil refineries themselves.[10]

The Depression sparked renewed interest in ethyl alcohol blends, particularly in the Midwest, where corn prices had dropped to 22 percent of their pre-Depression levels. The oil industry fought tooth and nail to discredit the practice as corn alcohol blends were marketed under names like "Vegaline" and "Alcolene." Playing upon the prohibition-era stigma of alcoholic beverages, the petroleum industry sullied the reputation of alcohol fuel and everything associated with it, portraying its supporters as crackpots and bootleggers. The Earl Coryell Company of Lincoln, Nebraska began selling "Corn Alcohol Gasoline" in the early 1930s, only to be undercut in its market by Ethyl Corporation, which refused to sell its product to anyone who sold ethanol blends. An antitrust lawsuit settled by the Supreme Court in 1940 found against Ethyl Corp., but by that time, tetraethyl lead was well entrenched in the marketplace.

At the same time, a broader movement was afoot to come up with new uses for farm products, including corn ethanol. Scientists, farmers and industrialists — among them Henry Ford — felt that using soy, hemp, corn and crop biomass for innovative manufacturing purposes would provide a new market for agricultural products and create jobs for rural Americans. Dubbed "farm chemurgy," it got national attention when a Ford-sponsored conference at Dearborn in 1935 spawned the National Farm Chemurgic Council, a trade group that survived for 40 years.

NEBRASKA STATE HISTORICAL SOCIETY

Fig. 1.9: *Nebraska's Governor Charles W. Bryan and Merrick County Sheriff J.J. Mohr fill up with corn ethanol-blended gas at a Coryell service station in Lincoln, Nebraska on April 11, 1933.*

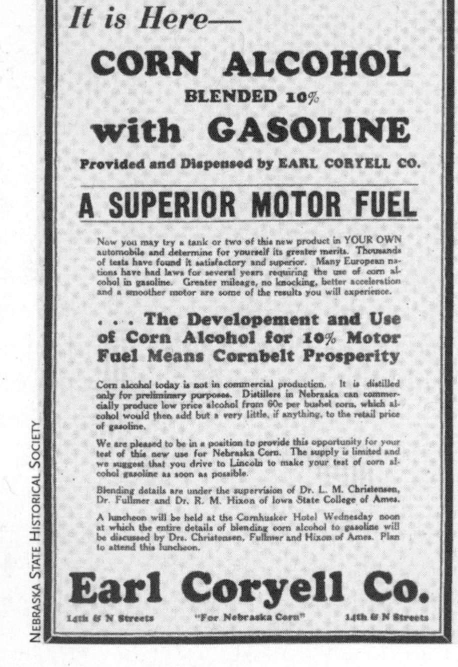

It is Here—

CORN ALCOHOL
BLENDED 10%
with GASOLINE

Provided and Dispensed by EARL CORYELL CO.

A SUPERIOR MOTOR FUEL

Now you may try a tank or two of this new product in YOUR OWN automobile and determine for yourself its greater merits. Thousands of tests have found it satisfactory and superior. Many European nations have had laws for several years requiring the use of corn alcohol in gasoline. Greater mileage, no knocking, better acceleration and a smoother motor are some of the results you will experience.

. . . The Developement and Use of Corn Alcohol for 10% Motor Fuel Means Cornbelt Prosperity

Corn alcohol today is not in commercial production. It is distilled only for preliminary purposes. Distillers in Nebraska can commercially produce low price alcohol from 60c per bushel corn, which alcohol would then add but a very little, if anything, to the retail price of gasoline.

We are pleased to be in a position to provide this opportunity for your test of this new use for Nebraska Corn. The supply is limited and we suggest that you drive to Lincoln to make your test of corn alcohol gasoline as soon as possible.

Blending details are under the supervision of Dr. L. M. Christensen, Dr. Fullmer and Dr. R. M. Hixon of Iowa State College of Ames.

A luncheon will be held at the Cornhusker Hotel Wednesday noon at which the entire details of blending corn alcohol to gasoline will be discussed by Drs. Christensen, Fullmer and Hixon of Ames. Plan to attend this luncheon.

Earl Coryell Co.

14th & N Streets | "For Nebraska Corn" | 14th & N Streets

NEBRASKA STATE HISTORICAL SOCIETY

Fig. 1.10: *A newspaper advertisement for the Earl Coryell Company touts the benefits of corn ethanol fuel in the heart of the depression-era corn belt.*

In 1937, the Agrol plant in Atchison, Kansas, financed by the Chemical Foundation, Inc. (an administrator of alien properties and patents seized after World War I) began marketing an ethyl alcohol gasoline blend throughout the Midwest. Priced competitively with leaded gasoline at retail, it was still more costly at the wholesale level, which eventually affected market sales. Though the Atchison plant closed in 1939, the experience gained there prepared chemists and engineers to manufacture the vast quantities of ethyl alcohol necessary to supply the war effort with torpedo and aviation fuel, and the crucial synthetic rubber, styrene-butadiene.

The post-war period ushered in an economy highly dependent upon cheap petroleum and unprecedented growth. Because of the low cost of oil, ethanol was all but non-existent as a commercial fuel, and didn't return to the public eye until the mid-1970s, when international events prompted a fresh look at domestic fuel resources.

Gasohol and E-85

The oil crisis of 1973-74, spurred by the decision of the Organization of Arab Petroleum Exporting Countries (OAPEC) to cease oil shipments to countries that supported Israel in the October 1973 Yom Kippur War, had immediate effects on the US economy. The embargo was lifted in March of 1974, but driven by the collusion of the Organization of Petroleum Exporting Countries (OPEC) to raise world oil prices, the market price of crude oil quadrupled in a matter of months.

A number of legislative actions prompted immediate research into the use of cellulose and biomass for the manufacture of alcohol fuels, and corn-producing states lobbied for ethanol-gasoline blends to reduce the consumption of imported petroleum. The early efforts to market ethanol-gasoline blends were met with resistance from wholesalers and consumers alike, but the Energy Tax Act of 1978 helped the burgeoning movement by recognizing and defining gasohol as "a blend of gasoline with at least 10 percent alcohol by volume, excluding alcohol made from petroleum, natural gas or coal," [11] effectively mandating renewable feedstocks and subsidizing the use of alcohol fuel with that portion of the federal excise tax on gasoline.

By the arrival of the second oil crisis in 1979, commercial alcohol-blended fuels were being sold by Chevron, Amoco, Texaco and others. Retail sales of gasohol were publicized with labeling on individual fuel dispensers and the E-10 designation in some markets. Simultaneously, as the US began to phase out lead in gasoline (a complete ban was set for 1986) ethanol became more appealing as an octane booster. At that time, only a handful of commercial ethanol producers existed, manufacturing about 50 million gallons of fuel alcohol per year.

After a series of federal subsidy increases and a tariff on imported ethanol, domestic ethanol output increased to nearly 600 million gallons per year by the mid-1980s. Even as the price of crude oil dropped, making ethanol less cost-competitive, environmental concerns over carbon monoxide levels began to focus attention on the use of oxygenated fuels. Ethanol was a viable oxygenate, as was methyl tertiary butyl ether (MTBE) and ethyl tertiary butyl ether (ETBE). Of the latter two, only ETBE used ethanol in its manufacture; MTBE was made from natural gas and petroleum, and dominated the market because of its lower cost.

By 1992, the mandates of the Clean Air Act to use oxygenated fuels took effect in 39 major carbon monoxide "non-attainment" areas, followed three years later by nine severe ozone non-attainment areas. Gasoline was reformulated in many urban areas throughout the country to include oxygenates, chiefly MTBE. However, by the late 1990s, traces of MTBE were starting to show up in drinking water sources, and some states began to prohibit the use of MTBE in motor fuels, shifting demand to ethanol and ETBE. In 2000, the Environmental Protection Agency recommended a national phase-out of MTBE, but states that were major users of the oxygenate were moving in that direction regardless. Before long, major refineries voluntarily chose to use ethanol-based oxygenates in their reformulated gasoline, fearing eventual litigation over MTBE's carcinogenic effects.

The establishment of E-85 as a separate gasoline blend began in the early 1990s with the Energy Policy Act, which among other things provided for an alternative transportation fuel comprised of 85 percent ethanol and 15 percent gasoline. Major auto manufacturers in 1997 began mass-producing so-called "Flex Fuel" vehicles designed to use unleaded gasoline, E-85 or any combination of the two.

RICHARD FREUDENBERGER

Fig. 1.11: *An E-85 fuel pump, set up to dispense 85 percent ethanol fuel for Flex Fuel vehicles.*

At this writing, 6.8 million Flex Fuel vehicles are on the road, including cars, minivans and pickup trucks. The Flex Fuel vehicles are priced competitively with their gasoline counterparts, and are factory warranted and EPA approved. The main difference between a Flex Fuel vehicle and a conventional gasoline vehicle is the addition of a fuel sensor that detects the ethanol/gasoline ratio. Other components such as the fuel tank, fuel lines and fuel injectors have been modified slightly. The computer chip in the on-board electronic control module/powertrain control module (ECM/PCM) is reprogrammed to broaden its control range for the alternative fuel.

Flex Fuel vehicles can experience a reduction in fuel mileage when using E-85, because the engines are not optimized exclusively for alcohol fuel (see more on that in Chapter 9). Generally, E-85 is priced below unleaded regular to compensate for the difference. Because it's made from agricultural products, it is considered a renewable fuel, and has excellent anti-knock qualities owing to its 105-octane rating. From an environmental standpoint, E-85 reduces lifecycle greenhouse gas emissions by approximately 20 percent compared to unleaded gasoline, and it further reduces evaporative and carbon monoxide emissions and toxic emissions such as benzene and other known carcinogens.

E-85 contains 15 percent unleaded gasoline to allow the Flex Fuel vehicles to start in cool

RICHARD FREUDENBERGER

Fig. 1.12: *The author's 2000 Ford Ranger Flex Fuel vehicle appears no different than the standard gasoline model. The fuel door is labeled for E-85 and regular unleaded.*

conditions. In particularly cold climates, winter E-85 may actually contain a slightly higher level of gasoline to further aid in cold starting, though it's still labeled E-85. Some ethanol fuel advocates maintain that the entire E-85 program is simply a concession to auto manufacturers and the petroleum industry to circumvent the modest additional cost of a factory-equipped auxiliary cold-starting system. They contend that it would have been just as practicable to manufacture E-100 (pure ethanol) fuel and Flex Fuel vehicles capable of operating on any combination of pure ethanol and gasoline, as Brazil does.

Brazil's Booming Ethanol Program

The US, of course, was not the only country affected by the inconsistencies of the world petroleum markets. Brazil, with significant domestic oil resources but limited infrastructure to develop it, relied heavily on petroleum imports and was hard-hit by the 1973 embargo. With a serious trade deficit at hand and the threat of monumental inflation, the government launched its "Programa Nacional do Alcool" or "Pro-Alcohol" program. Its goal was the phase-out of all automotive fuel derived from fossil sources, to be replaced with domestically produced ethanol.

Initially, anhydrous (water-free) alcohol was blended with gasoline in a mandatory 20/80 percent mix the equivalent of E-20. In 1979, after the second oil crisis, the program expanded to include straight-alcohol vehicles. At the time, there were no commercially available Flex Fuel vehicles. The choices were either E-20 gasohol cars (with minor adjustments made by the manufacturers, owners or by private mechanics) or E-100 vehicles, specially manufactured to run on the hydrous ethanol

Richard Freudenberger

Fig. 1.13: *A General Motors vehicle in Brazil displays the alcohol fuel label on its trunk lid.*

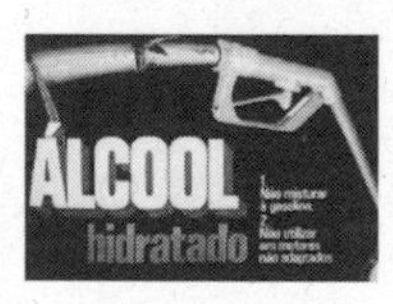

Instituto de Economia Agricola

Fig. 1.14: *A sign designating hydrated alcohol at a Brazilian fuel pump.*

sold at service stations. (Hydrous, or azeotropic, ethanol is a 96-percent pure alcohol containing 4 percent water, the highest concentration obtainable in distillation without going through additional processes; water at that amount is not detrimental to the fuel. Anhydrous alcohol is only required when it's mixed with gasoline, after which it does not separate out.)

When I went to Brazil in 1980 to study the national alcohol program and speak with the auto manufacturers' engineers, the program was in the beginning phases of a huge growth curve. Close to 240,000 E-100 cars would be made that year, representing about 25 percent of the total manufactured. To be truthful, the program was still a social experiment of sorts, and not all the motoring public embraced the changes enthusiastically.

Nonetheless, the Brazilian government set the stage for success by planning carefully. It guaranteed purchases of ethanol by Petrobras, the government-owned oil company. Gasoline and ethanol prices were fixed by the state, with straight ethanol priced at some 60 percent of the retail gasoline price. E-100 car fleets were put into service by public utilities such as the telephone and power companies, and low-interest loans were extended to agricultural and industrial companies engaged in the business of ethanol manufacturing. To this day, Brazilian ethanol production is unsubsidized and does not benefit from import tariffs.

The Pro-Alcohol program was based on sugarcane because of its availability, low cost and favorable energy content. The cane's sugar content comprises slightly more than 30 percent of the energy available in the plant. About 35 percent is in the leaves and tips, which remain in the fields, and the balance of 35 percent or so is in the fibrous bagasse left over after the cane is pressed. Consequently, ethanol production from sugar cane is a nearly self-contained system, with a portion of the bagasse burned as fuel at the processing mill to provide heat for distillation and for the production of electricity. Enough is left over that surplus power — about 15 percent of the total — can be sold to utilities, which accounts for three percent of the nation's electrical generation.[12]

Only one percent of Brazil's arable land is used to produce ethanol, yet it provides fuel for nearly 50 percent of the automobiles in the country. Over five billion gallons of fuel ethanol was produced in 2007, and Brazil remains the largest exporter of ethanol in the world, shipping nearly 20 percent of its production to world markets. Because sugarcane is an efficient photosynthesizer, it's very effective at converting sunlight to energy, and ethanol yields per acre can easily be double or triple that of corn — or more, with multiple harvests. Moreover, sugarcane is already in the saccharin form that yeasts require for fermentation, so the conversion from starch to sugar is unnecessary. Sugarcane ethanol's energy balance — the ratio of energy obtained from the ethanol versus the energy used to produce it — is extremely favorable, averaging 7 to 8 times greater compared to corn's 1.3 times. As a result, ethanol can be sold in Brazil for a price that is on average one dollar below that of gasoline.[13] It's no wonder that virtually all the fuel stations in Brazil offer E-100 as compared to one percent of the stations in the US.

Since 2003, Brazilian automakers have been manufacturing Flex Fuel vehicles that run on hydrous alcohol, an E-20 or E-25 gasoline blend, or any combination in between. Fiat, Volkswagen, General Motors, Ford and most of the major Japanese manufacturers market the Flex Fuel configuration, which accounted for 87 percent of Brazil's car sales by the middle of 2008.[14] Currently, Flex Fuel models include a small auxiliary gasoline reservoir to resolve cold-starting problems for vehicles used in Brazil's cooler southern states, but beginning in 2009, improved technology will eliminate the need for the supplementary tank altogether.

Since its inception in the late 1970s, the Pro-Alcohol program has cut the number of Brazilian vehicles running solely on gasoline by 12.3 million.[15] The county's demand was so great that for a period in the 1990s, there was not enough ethanol available to supply the market, a situation that was resolved with increased commitment despite low petroleum

Fiat Automoveis

Fig. 1.15: *The new Fiat Siena ELX 1.4 Tetrafuel can operate on gasoline, alcohol, compressed natural gas or a combination of fuels*

prices at the time. In addition, the notorious air quality of cities such as São Paulo has improved with the introduction of ethanol-fueled vehicles, and the reduction of greenhouse gases nationwide credited to the use of ethanol is estimated at 86 percent.[16]

The Economics of Ethanol

In 2008, reports from the World Bank, Oxfam and a number of environmental and humanitarian groups criticized ethanol as the solution to neither the oil crisis nor the climate crisis, while contributing to the rise in food prices. The details paint a picture of unfair trade policies, the conversion of forests to new biofuel cropland, errant government subsidies, and negative energy balances. But people within the ethanol industry, and many economists and global energy experts hold a different view. They say that the cost of energy impacts every food product in the market. With the rising price of petroleum, every aspect of food production is affected. It takes energy to plant, harvest, process, package and ship everything we eat. Labor and transportation are far more influential than corn price. The impact of the price of corn is only a small fraction of the total Consumer Price Index — the price of energy and increasing petroleum prices have twice the effect as equivalent corn price rises. Furthermore, US ethanol is produced from field corn, which is not widely used for human consumption. It's primarily a livestock feed, and the majority of corn exports go to developed countries, not impoverished nations. What's more important is that the grain itself is not "used up" in the manufacture of ethanol. Alcohol production uses only the starch portion of the kernel, while the remaining protein, minerals, vitamins and fiber is sold as valuable livestock feed in the form of distiller's grains. This "digestible energy" is highly marketable. For each 56-pound bushel of corn that goes into the ethanol process, 18 pounds of Dried Distiller's Grains with Solubles (DDGS) is recovered. The Dried Distiller's Grains have more feed value than the equivalent amount of corn, averaging about 26 percent protein, an 18 percent increase over corn alone.[17]

Food vs. Fuel

None of the anxiety over rising corn prices escaped the attention of the nation's food industry, whose leaders expressed concerns over diverting grains to fuel use and the subsequent effect it might have on food staples and livestock feed. True to form, costs of grain-based products rose noticeably, a boon for price-starved farmers, but a bust for consumers of everything from meat to soda pop to dog food — anything that uses corn sweeteners in its manufacture.

Moreover, the demand for corn has spurred a large agricultural interest in production. (About 20 percent of the corn crop goes into the manufacture of ethanol.) Since corn is an especially nitrogen-fertilizer intensive crop, the inevitable runoff from farm fields has been linked to algae blooms in the nation's waterways, some widespread and intensive to the point of affecting marine life and subsequent seafood harvests.

On a larger scale, the demand for corn and other biofuel crops has prompted further agricultural development globally, with land-clearing, deforestation and unprecedented water usage the result.

Despite President Bush's State-of-the-Union appeal, the most expedient way to make ethanol at this point in time is with corn, not with wood chips or switchgrass as he envisioned. But there's a caveat: All of this involves massive market-scale production — manufacture on a level so enormous as to apply only to entities entrenched in the world of Agribusiness. At this point, making ethanol is so profitable — thanks to government subsidies and unusually high oil prices — that ethanol plants can't be built quickly enough to meet demand. And that will continue as long as fuel prices are high and corn prices remain relatively low.

Yet what many people are talking about is far removed from this. It is fuel production on an appropriate local level, either cooperatively or within a farm system that has access to a variety of reliable feedstocks. Chapter 5 discusses raw materials and the value of starches and sugars in depth, but it's important to realize early on that what's good for the goose isn't always good for the gander. At some point, the price of petroleum, or the cost of refining it or delivering it, may be so high as to make small-scale production of farm ethanol very attractive to those who seek to be in control of their own fortunes. Going one step further, acquiring the raw materials for fuel from a dependable local source — be it a product of the area economy, a crop chosen for its conversion value, or agricultural waste — offers a degree of separation from an uncertain and perhaps unstable national economy, which holds a certain amount of appeal and even comfort.

There was a time, well within the memories of some folks living today, when the energy used to run a farm or homestead came from the farm itself. The food was grown there, the wood that cooked the food and heated the house was felled there, and the muscle that planted and harvested the crops was fed on the hay and grasses raised there. Wind pumped the water, and often provided basic electrical needs. Today's technology can only improve on that.

Things may yet come full circle for those who plan. The ethanol made from farm crops can power the tractors, fuel the furnaces, run the generators and feed the livestock with protein-rich distiller's grains. By-products from ethanol production can fire the cookers that boil the mash and heat the distillation columns. And the labor can remain on the farm and in the community it supports, rather than left to the uncertainties of an unsure outside market.

TWO
What You'll Need to Start

Just curious or serious? • Gauging your commitment • Long-term objectives

One of the wonderful things about biofuel is that you can make it yourself. Fuel made from local resources makes it possible, at last, to get free of the economic shackles that bind us to a system that has become so volatile as to impact our own financial and societal well being, even down to the individual level. An entire economy was built on cheap petroleum, and the dearth of that cheap petroleum now threatens to destroy it.

So it's only natural that the subject of making one's own fuel would pique the curiosity even of those not normally inclined toward self-reliance. When biodiesel (an "organic" diesel fuel made from vegetable oil, methanol and a potassium hydroxide catalyst) came upon the scene, it took no time at all for backyard biodiesel operations to spring up — micro-enterprises operating out of garages and outbuildings, cranking out just enough fuel to feed the family car or run the truck and tractor. Since the initial investment for this type of endeavor is small (it doesn't require much more than a recycled water heater, a storage drum or two, and some plumbing), average folks — separate from the far-seeing entrepreneurial souls who had plans to market the brew — came to the technology in significant numbers, and today there is a healthy grass-roots following that has taken up biodiesel as a matter of course.

Alcohol fuel is different in a number of ways. Yes, it is also a home-grown fuel. And yes, it can be manufactured on a very small scale using salvaged equipment, and without a lot of formal training and experience. But making ethanol economically is, like many enterprises, a matter of scale. Later on in this book I'll talk a bit about small stills and what you can expect of them. For now, I'll use a simple pot-boiler with a 2-inch column as an example. (Think of a 55-gallon drum with a laundry drainpipe rising from the top of it.) That little cooker will make slightly less than one gallon of liquid fuel per hour. To put that in perspective, if you're driving a Ford pickup (which for many years was the best-selling "car" in America), you'll consume *four* gallons of fuel in one hour at interstate speeds.

Naturally, you can always take the permaculturist's approach and look at the big picture — making more fuel isn't the only solution; driving less and doing so in smaller vehicles is another option. Or, you might fittingly be considering home-brewed fuel for your rototiller, garden tractor, or other lesser use. But if you're serious about making ethanol, you'll have to be prepared to make an investment not only of capital, but of time and effort as well. This, of course, is true with every homemade fuel.

Table 2.1 offers a brief summary of the comparisons between ethanol, biodiesel and straight vegetable oil (SVO), which is a third option that may be worth considering for some people if they're set on a substitute for petroleum. My purpose here isn't to digress; the point is to bring up some alternatives that might help you make the best decision for your situation. As you can see from Table 2.1, alcohol is a fairly flexible fuel so far as engines and equipment is concerned. It stores well and doesn't require commercial chemicals in its manufacture. It's also a product of raw feedstocks that are much more available than you might imagine, which I'll get to in Chapter 5.

Understanding Your Commitment

On a practical level, the best place to start would be in trying to establish how much fuel you'll actually need. If you *know* you'll only be experimenting, you can limit your investment to scrap and salvaged parts and probably

Table 2.1
Comparison Between Ethanol, Biodiesel and SVO Fuels

	Investment Cost	Learning Curve	Operating Cost	Tailpipe Emissions	Time Investment	Facility Requirements	Liability Exposure
Ethanol	Moderate to Substantial	Substantial	Moderate	Minimal	Moderate to Substantial	Moderate to Substantial	Moderate
Biodiesel	Minimal	Moderate	Moderate	Minimal to Moderate	Moderate	Moderate to Minimal	Moderate
SVO	Moderate	Minimal	Minimal	Insufficient Data	Moderate	Minimal	Minimal

Note: Due to variation of scale, the information here is somewhat subjective and is presented for comparison purposes only. Specific data for Tailpipe Emissions of Straight Vegetable Oil fuel is inconclusive mainly because of the difference in properties between various oils and the degree to which they are heated prior to combustion. In general, all three fuels demonstrate reduced levels of Hydrocarbon (HC), Carbon Monoxide (CO), and Particulate Matter (PM in diesel engine) emissions as compared to conventional fuels, but show increased levels of Nitrous Oxides (NOX) in varying degrees.

locate a free source of feedstock without much difficulty (more on that in Chapter 5 as well). This approach may be a good initiation to ethanol fuel for everyone, since it'll give a taste of what it takes to prepare a mash ferment, cook it, and run a batch of alcohol through a column. A few simple implements, a 55-gallon drum, and a small batch still will suffice for this purpose.

CLARENCE GOOSEN

Fig. 2.1: *A small boiler vat and distillation tower is sufficient for experimentation.*

But if you don't want to start building quite yet, it shouldn't be too hard to calculate what volume of fuel you might go through in a month. A monthly assessment is probably best for most people because it's a manageable period and a long enough chunk of time to allow some flexibility in the production process. If you use a credit card for gasoline purchases,

ICHARD FREUDENBERGER

Fig. 2.2: *A very simple pot distillation still with an external condenser barrel, confiscated in 1933 in Western North Carolina. An outer brick lining surrounded by wire mesh contained firebox heat and helped subdue potential explosions that might occur if the distillate "worm" tube became clogged with mash.*

DAN WEST

Fig. 2.3: Missouri fruit farmer Dan West has experimented with a solar thermal heat source to distill ethanol in this small test still.

CLARENCE GOOSEN

Fig. 2.4: A 6-inch still, built on a mobile trailer for demonstration sessions. Many of the components for a still of this size can be salvaged or recycled from other agricultural or commercial applications.

simply check your receipts. Otherwise, you can keep a mileage log and estimate the fuel mileage for each car or truck that you drive. Don't rely on the EPA estimated fuel mileage figures for your vehicles because they represent a specific test cycle developed for the EPA and are useful more as a point of reference between vehicles in the same class than as an actual real-world mileage measure.

Once you've determined your fuel need for a fixed period, you can look at equipment. Chapter 8 gets into the cookers, stills, and handling the equipment you'll need for various sized operations, but a simple summary is presented in Chapter 7 (see Table 7.5). There, you'll see that a 6-inch column still has a theoretical capacity of six gallons of fuel-grade (190-proof) ethanol per hour. (This is an average, because a host of factors determines output, including the design of the column, the heat source, and how well temperatures are controlled; actual output can reach ten gallons per hour under ideal conditions.) An 8-inch column can yield 10 to 18 gallons per hour. Going up to a 12-inch column boosts output to 24 gallons per hour, with up to 35 gph possible. But upgrading comes at a cost in both equipment and time, and at some point, you may have to consider the distillery process you'll be using as well.

In a batch operation — in which you prepare and ferment a feedstock and then distill it until the batch is exhausted — the fermented mash is sent through the still as soon as possible to reduce the chance of bacteria infecting the ferment. The mash contains about 10 to 12 percent alcohol, which is separated from the rest of the liquid in the distillation process. With this procedure, it's usually most convenient to set aside a block of time each month to go through the production steps and then store the alcohol you produce until it's time for the next run. A batch operation can be a relatively low-tech endeavor and the components can be compiled from salvaged or fabricated parts, recycled or new off-the-shelf hardware, or any combination thereof.

The other approach is called continuous-feed distillation, in which the process proceeds, for lack of a better word, continuously. The fermented mash is introduced to the distillation column and recycled until all the alcohol is recovered; fresh mash is fed in at the same rate that alcohol is drawn out. The feedstock is prepared and fermented on a rotating basis so that when one lot is ready for distillation, a fresh batch is just beginning to ferment, and other batches are working at various stages in between. This approach requires more equipment and a lot more control. The investment is greater because the still's construction is more sophisticated and it requires complex monitors and controls. And it needs a dedicated manager, either a human or a computer program.

Practically speaking, most people involved in small-scale alcohol production will stick with a batch process. When production levels reach 35 gallons per hour or more, it's worth some investigation into whether a continuous process might be feasible, with some additional investment.

At some point, if your alcohol production levels exceed the needs of your vehicles or equipment, you can always consider the sale of alcohol for profit. This puts your operation on a whole new plane, and it requires additional permitting, insurance, a surety bond, environmental controls, revenue taxes, and a firm commitment to business practices. Most small-scale operators and probably very few casual producers will have any interest in going to this level.

There are, however, benefits to using, reselling or bartering some of the co-products of the distillation process to offset the costs of your production. Even in a small-scale operation, these could include Distiller's Dried Grains (a valuable animal feed commodity), carbon dioxide (a gas by-product of fermentation used in bottling, packaging and greenhouse operations), fertilizers and amendments (depending on the feedstock used), and silage. In addition, there are some collateral by-products that could be developed on a profit basis for specialty markets. Yeast culturing, enzyme and malt production, and the processing of biomass fuels are just some of the possibilities.

CLARENCE GOOSEN

Fig. 2.5: *Solid fuel — in this case wood — heats transfer oil in a specially made boiler to provide thermal input for an early 6-inch still design.*

Short-Term or the Long Haul?

For a small-scale producer, the urge may be great to just jump in and start making alcohol, but I would advise against being impulsive. Even an experimental approach takes some planning, and a well-advised first step is to initiate the application process for your federal Alcohol

Fuel Producer's permit, which I describe in the next chapter, along with some commentary on state requirements and information on insurance and other details. Securing the permits will not only make your enterprise legal, but in going through the permitting process, you will be compelled to think through potential problems and make the best decisions, which is best for you in the long run.

Making alcohol requires an endeavor into multiple disciplines. It is ideal for the hands-on Renaissance man or woman who has the patience to learn new things and acquire disparate skills. A successful enterprise, taken from start to finish, will ask that you become a farmer, chemist, welder, engineer, mechanic and entrepreneur all in one. For some, it's the perfect job. Others may just want to get into the business of alcohol production rather than the guts of it. Either way, it's best to approach the process one step at a time with your eyes wide open, and let experience steer you in the best direction.

THREE

Federal and State Requirements

Alcohol Fuel Producer permit • State and local jurisdiction • Insurance and bonding • Environmental impact • Fire code • Plant premises and storage • Denaturing process • Proof gallon measurement • Essential record keeping

For reasons that should be apparent, federal, state, and even local authorities have a keen interest in the manufacture of fuel alcohol. Their concern relates in part to the potential for evasion and abuse of the intricate and long-standing tax structure that exists for the manufacture of distilled spirits for human consumption. (Other concerns include environmental impact and fire safety on a local level, which I'll speak to shortly.)

The federal government's involvement with distilled beverages and taxation can arguably be traced to the Whiskey Rebellion, or Insurrection, of 1794, in which farmers from Western Pennsylvania took exception to federal laws enacted three years prior that imposed a per-gallon tax on home-produced spirits, and a flat levy on larger alcohol producers.

Then Secretary of the Treasury Alexander Hamilton urged Congress to impose the tax, both to build up treasury reserves and to impose some degree of social discipline on the population of the burgeoning westernmost counties. Farmers and settlers, on the other hand, had traditionally converted their excess grains to spirits — it was a convenient way to bring the grain to market — and strongly opposed what they considered an unjust tax. Moreover, distilled spirits were commonly used in lieu of currency for barter in frontier communities from Pennsylvania to Georgia, where cash money was scarce.

The insurrection led to the raising of a sizeable militia, which eventually brought a small group of men to justice, and, not coincidentally, established the federal government's authority over states' citizens. The tax was repealed 12 years after it was enacted, but not before encouraging distillers to practice their skills in the frontier regions of Tennessee and Kentucky, well beyond the reach of federal influence.

Since 1862, levies have been raised by the federal government in the form of excise taxes. The rate of $0.20 per 100-proof gallon established during the Civil War is currently set at $13.50, where it's been since 1991. (A proof-gallon is defined as one liquid gallon of spirits that is 50 percent alcohol at 60°F.) Thirty-two state governments establish their own alcohol excise taxes, ranging from $1.50 per proof-gallon in Maryland and the District of Columbia to $12.80 in Alaska. The remaining 18 states simply mark up liquor prices. An indication of how profitable these taxes can be is reflected in California's $3.30 per gallon rate, which generates about $138 million in revenue annually.

Enter Fuel Alcohol

Alcohol fuel has piqued the public's interest — along with the interest of the farm lobby — through several periods in the last century. The repeal of the tax on non-beverage ethanol use in 1906, a number of Depression-era fuel programs for corn farmers, and the "home-grown" response to the energy crisis of the mid-1970s all came about as ethyl alcohol was viewed anew as a potential motor fuel. Once the automobile established itself on the American scene, it became clear that alcohol as a fuel was viewed as a separate matter from its use as a beverage, and as such, excise taxes did not apply.

In 1954, the Bureau of Alcohol, Tobacco and Firearms (BATF) included language in the Internal Revenue Code that allowed individuals and businesses to qualify for Experimental Distilled Spirits permits, which included the manufacture of ethyl alcohol for use as fuel.[1] Once denatured, or rendered unfit for beverage use — typically with the addition of at least two percent unleaded gasoline, kerosene or naptha — the distilled liquid can be used off-premises as a fuel. (Interestingly enough, if the distilled spirits were used on-premises as a fuel, or on the premises of a separate farm or location described in the original permit application, no denaturing was required, even for transport between the manufacturing plant and the remote site.)

As a result of the Crude Oil Windfall Profit Tax Act of 1980, amendments were made to the US Tax Code to include a section that authorized "the establishment of a distilled spirits plant solely for the purpose of producing, processing and storing, and using or distributing, distilled spirits to be used exclusively for fuel use."[2] In essence, these changes significantly simplified the requirements for maintaining a production facility exclusively for fuel, and they represent the code currently in use by the Alcohol and Tobacco Tax and Trade Bureau. (In January 2003 the Homeland Security Act split the functions of the original organization so that enforcement came under the authority of the Department of Justice, and trade and revenue fell under the Treasury Department's Alcohol and Tobacco Tax and Trade Bureau, or TTB.)

Three levels of Alcohol Fuel Producer (AFP) now exist, appropriately labeled Small, Medium and Large. For most of us, only the Small AFP permit applies, as it allows the manufacture of up to 10,000 proof-gallons of alcohol per calendar year, a quantity adequate to operate several vehicles.

In practice, defining a proof-gallon as one liquid gallon of spirits that is 50 percent alcohol at 60°F provides a common denominator for measuring volume for ethanol of various strengths. For example, 50 gallons of 190-proof ethanol equals 95 proof gallons. The formula is expressed as so: proof strength of the liquid × volume in gallons ÷ 100 = proof gallons. In TTB terms, a gallon is described as a "wine gallon," or 128 ounces of ethanol (231 cubic inches in volume), regardless of strength.

Alcohol Fuel Producer Permit

In January of 1979, when the original *Mother Earth News* published the story of a Minnesota farmer's foray into alcohol fuel production as a way to deal with the high cost of fuel, federal requirements for making ethanol were considerably more complex than they are today. Though there had always been regulations for commercial alcohol producers, a new breed of individuals was more interested in the self-sufficiency aspects of alcohol as a fuel than in its commercial value.

Such people found support in an obscure section of Title 27, Code of Federal Regulations that allowed for the establishment of an "Experimental Distilled Spirits Plant" to experiment in or to develop, among other things, new feedstock sources and processes for making and refining spirits. The 25-year-old legislation really wasn't tailored to fuel producers, but was meant to deal with testing and research for industrial uses; nonetheless, the BATF — recognizing the significance of a domestic fuel source in the midst of a petroleum crisis — worked to approve experimental plants in its seven regional offices across the country.

The permit for an experimental plant had no formal application form. Applicants were asked to describe the purpose of the operation, their intentions, and to provide a description of the premises and equipment, security measures, by-product use, rate of production, and duration of operations. Plant operators had to denature the product or pay tax on it, post an indemnity and distilled spirits bond, provide power of attorney, supply consent from property owners, allow access to manufacturing records, and submit environmental impact statements. Compliance with all state and local requirements was the responsibility of the applicant, and BATF authorization could be rescinded at any time for revenue or administrative infractions. The product could not be sold or given away, and the permit had a maximum two-year duration. In truth, the Title 27 law made little distinction between beverage alcohol and fuel alcohol, but it was the only one that could be applied at the time.

By July 1980, the Bureau had simplified matters for fuel producers by introducing a new type of permit that removed some of the obstacles to production. Principally, these were the elimination of the bond for small plants, the establishment of simpler application and record-keeping procedures, and permitting of commercial for-profit operations. The permit allowed for Small, Medium and Large Plants, each categorized by annual proof-gallon production (respectively, 10,000; 10,000–500,000;

and more than 500,000). The Small plant category, as mentioned earlier, is likely all that most individual producers would need, as it permits manufacture of 5,263 gallons of 190-proof alcohol fuel per year. (A 25 mile-per- gallon vehicle would use about 600 gallons driving 15,000 miles annually.)

The main federal form is TTB F 5110.74, entitled "Application and Permit for an Alcohol Fuel Producer Under 26 U.S.C. 5181." You can get a copy of the form by calling (513) 684-7150 or (877) 882-3277, or it can be downloaded at www.ttb.gov/forms/f511074.pdf (or, on the forms page, go to "TTB: Other Alcohol," then "Application: AFP Under 26 U.S.C. 5181.") Only two restrictions apply immediately, and they concern applicants who are citizens of, or associated with, a foreign country (or have lived as an adult in a foreign country for more than two years), or those who plan to use a premises that is eligible for the National Register of Historic Places. Additional forms are required for those situations.

Otherwise, the form is clear and straight-up. The application includes a cover sheet with instructions for completing the form, and the address and contact information for each of the seven TTB regional offices (which one you need to contact depends on the state in which your operation will conduct its business).

When filling out the application, it would be prudent to think about how you intend to use your operation, and to cover all future possibilities accordingly. For example, Item 1 asks you to declare the type of plant you wish to set up. A Small plant of 10,000 proof-gallons or less needs no bond; the larger two require that you put up a Distilled Spirits Bond (which you can get through your insurance agent or from a surety company). Since you can always upgrade simply by filing an amended permit and furnishing a bond, start small and save yourself unnecessary bother at the beginning.

Item 7 asks for the physical location of the alcohol plant, with a street or rural route number. Your mailing address goes at Item 8, and may be different from the plant's location. Items 9 and 10 ask for information on the premises, because if it's leased or otherwise not owned by the applicant, the TTB and state and local officers will need signed permission from the owner for access, to avoid a potential trespassing charge.

The balance of the application concerns materials, equipment and the physical site. Item 11a is for the still manufacturer. If you bought the column, list the maker; if you made it yourself, you can simply put "owner." Likewise for Item 11b, which requests a serial number. A manufactured still should have a number; you can assign your own to owner-built equipment. Unless you know for certain that your still is continuous feed or some unusual design, for Item 11c you can list it as a batch still with a packed column, or if it's a plate type, with a perforated plate column. The capacity, Item 11d, is to be given in proof-gallons distillable in a 24-hour period. This information should be supported with more detailed data such as column diameter, number of plates or type of packing material, and the volume of the boiler vessel; the idea is to give the inspectors a way to calculate how much ethanol could actually be produced in

DEPARTMENT OF THE TREASURY
BUREAU OF ALCOHOL, TOBACCO AND FIREARMS

APPLICATION FOR AN ALCOHOL FUEL PRODUCER PERMIT

(Prepare in Triplicate. See Instructions)

ATF USE ONLY	
DATE RECEIVED	DATE RETURNED AFTER CORRECTIONS
PERMIT NUMBER	EFFECTIVE DATE

1. TYPE OF PLANT *(Check applicable box)*
(Complete for Original Application or when level of operation changes)
☐ SMALL - 10,000 Proof Gallons or Less*
☐ MEDIUM - More than 10,000 Proof Gallons but not more than 500,000*
☐ LARGE - More than 500,000 Proof Gallons*
*Proof Gallons to be produced and received during one calendar year *(See Instruction 5.)*

2. AMENDED PERMIT *(Check applicable box(es))*
(Change In)
☐ LEVEL OF OPERATIONS *(Increased operations by small and medium plants only.)*
From ______
To ______
(Change In)
☐ NAME OF PROPRIETOR
☐ LOCATION OF PLANT
☐ OTHER *(Explain)*
PERMIT NO. AFP- | STATE

3. NAME OF OWNER *(If partnership, include name of each partner)*

4. DAYTIME TELEPHONE NUMBER *(Include area code and extension)*

5. SOCIAL SECURITY NUMBER | 6. DATE OF BIRTH

7. LOCATION *(If no street address show rural route)*

8. MAILING ADDRESS *(If different from plant location) (RFD) or Street No., City, State, ZIP Code)*

9. PREMISES FOR ALCOHOL FUEL PLANT ARE *(Check applicable box)*
☐ OWNED BY THE APPLICANT *(Skip Item 10, go to Item 11)*
☐ NOT OWNED BY THE APPLICANT *(Complete Item 10)*

10. Officers of the Bureau of Alcohol, Tobacco and Firearms, State and local Officers, are granted access to the premises described by this application for an Alcohol Fuel Producer's Permit.
NAME OF PROPERTY OWNER
SIGNATURE OF/FOR PROPERTY OWNER | DATE

11. STILLS FOR FUEL PRODUCTION ON PLANT PREMISES

STILL MANUFACTURER *(If owner is the manufacturer write "Owner")* (a)	SERIAL NUMBER OF STILL (b)	KIND OF STILL *(Charge, Chamber, Continuous Still, or other (Specify))* (c)	CAPACITY *(Proof Gallons) (See Instruction 7.)* (d)

12. BASIC MATERIALS *(Other than yeasts or enzymes)* TO BE USED IN PRODUCTION OF SPIRITS *(Check applicable box(es))*
☐ GRAIN *(Corn, Wheat, Sorghum, Barley, etc.)* OR STARCH PRODUCTS *(Potatoes, Sweet Potatoes, etc.)*
☐ SUGAR BASED CROPS OR PRODUCTS *(Cane Sugar, Sugar Beets, Molasses, Sweet Sorghum, Beet Fodder, etc.)*
☐ FRUITS OR FRUIT PRODUCTS *(Grapes, Peaches, Apples, etc.)*
☐ FORAGE CROPS *(Alfalfa, Sudan Grass, Forage Sorghum, etc.)*
☐ CROP RESIDUE *(Garbage or other refuse)*
☐ OTHER *(Specify)* ______

13. DESCRIPTION OF SECURITY MEASURES *(Such as use of locks, fences, building alarms, etc.)* TO PROTECT PREMISES, CONTAINER(S), STILL(S) AND BUILDING(S) WHERE SPIRITS ARE STORED

REMOVE AND TURN CARBONS OVER WHEN COMPLETING THE REVERSE SIDE

ATF F 5110.74 (6-93)

Fig. 3.1: *The first page of the federal application for an Alcohol Fuel Producer Permit (Form 5110.74).*

Alcohol and Tobacco Tax and Trade Bureau, US Department of the Treasury

Fig. 3.2: The second page of the federal application for an Alcohol Fuel Producer Permit (Form 5110.74).

14. DIAGRAM OF PLANT PREMISES *(In the space provided or by attached map or diagram, show the area to be included for the alcohol fuel plant. Identify roads, streams, lakes, railroads, buildings, and other structures or topographical features on the diagram. Show location(s) where alcohol fuel plant operations will occur. The diagram should be in sufficient detail to locate your operations and premises.) (See instruction 8 for sample diagram.)*

15. I WILL COMPLY WITH THE CLEAN WATER ACT *(33 U.S.C. 1341(a)). (Will not discharge into navigable waters of the U.S.)*

☐ YES ☐ NO

16. IF THIS APPLICATION IS APPROVED AND THE PERMIT IS ISSUED, I CONSENT TO THE DISCLOSURE OF THE NAME AND ADDRESS SHOWN ON THE APPLICATION IN AN ATF PUBLICATION, "ALCOHOL FUEL PRODUCERS", WHICH MAY BE DISTRIBUTED ON REQUEST TO THE GENERAL PUBLIC *(including media, business, civic, government agencies, and others).* UNDER 26 U.S.C. 6103 YOU HAVE A LEGAL RIGHT NOT TO GIVE THIS RELEASE.

☐ YES ☐ NO *(A no response will have no effect on the consideration given this application)*

17. MEDIUM AND LARGE ALCOHOL FUEL PLANT APPLICANTS MUST PREPARE AND ATTACH THE ADDITIONAL INFORMATION SPECIFIED IN INSTRUCTION 11.

Under the penalties of perjury, I declare that I have examined this application, including the documents submitted in support thereof or incorporated therein by reference, and, to the best of my knowledge and belief, it is true, correct, and complete.

18. SIGNATURE OF/FOR APPLICANT	19. TITLE *(Owner, Partner, Corporate Officer)*	20. DATE

STOP
MAKE NO FURTHER ENTRIES ON THIS FORM

BUREAU OF ALCOHOL, TOBACCO AND FIREARMS

ALCOHOL FUEL PRODUCER PERMIT
UNDER 26 U.S.C. 5181

1. EFFECTIVE DATE

2. PERMIT NUMBER

AFP -

Pursuant to the above application and subject to applicable law and regulations and to the conditions set forth below you are hereby authorized and permitted at the premises described in your application to produce, process and store, and use or distribute distilled spirits *(Not including distilled spirits produced from petroleum, natural gas or coal)* exclusively for fuel use. The quantity to be produced and received from other plants during the calendar year may not exceed the quantity stated in this application.

This permit is continuing, and will remain in force until suspended, revoked, voluntarily surrendered, or automatically terminated. This permit does not allow you to operate in violation of State or local laws.

THIS PERMIT IS NOT TRANSFERABLE. In the event of any lease, sale, or other transfer of the operations authorized, or of any other change in the ownership or control of such operations, this permit shall automatically terminate. *(See 27 CFR 19.181 AND 19.920)*

3. SIGNATURE OF REGIONAL DIRECTOR (COMPLIANCE), BUREAU OF ALCOHOL, TOBACCO AND FIREARMS

CONDITIONS

1. That the permittee in good faith complies with the provisions of Chapter 51 of Title 26 of the United States Code and regulations issued thereunder.
2. That the permittee has made no false statements as to any material fact in his application for this permit.
3. That the permittee discloses all of the material information required by law and regulation.
4. That the permittee shall not violate or conspire to violate any law of the United States relating to intoxicating liquor and shall not be convicted of any offense which is punishable under Title 26 of the United States Code as a felony or of any conspiracy to commit such an offense.
5. That all persons employed by the permittee in good faith observe and conform to all of the terms and conditions of this permit.
6. That the permittee engages in the operations authorized by this permit within a 2 year period.
7. This permit is conditioned on compliance by you with the Clean Water Act (33 U.S.C. 1341391(a)).

ATF F 5110.74 (6-93)

order to figure tax liability in case the product was not used legitimately.

Item 12 asks for information on feedstock materials, and you should check off everything you could possibly have access to, even if you don't have it at the moment. Grains, sugar crops, fruits and fruit products, forage crops, and crop residue are all listed, and are all fair game for fermentation. Don't ignore the "Other" category, for here you can list anything that will ferment or has potential sugar content, including dairy and candy co-products, food processing waste, cattails or woody materials.

Security measures are covered in Item 13, where you should list what reasonable protection you will be providing for your containers, still and storage facility, whether it be locks, fences, alarm systems or security cameras. Reasonable is the key word here — but be honest, because an occasional visit from a federal TTF inspector is a real possibility, and it will be immediately obvious whether you're in compliance with your permit statement or not.

Item 14 asks for a simple diagram of the plant premises, identifying roads, streams, buildings and topographical features which will help to locate the site and alcohol operation. Item 15 covers compliance with the Clean Water Act, essentially asking if your operation will discharge into navigable waters, and Item 16 gives you the option of public disclosure. (A "no" answer does not affect your eligibility for the application, but keeps your name and address confidential and off the public list of Alcohol Fuel Producers.)

Once you've signed and dated the application, it will be processed and returned in a timely manner. Individual owners can sign for themselves, but a permit submitted in the name of a corporation must be signed by an authorized corporate representative. Any party signing on behalf of the applicant should have a Power of Attorney or other authorization available with the application. It's important to know that alcohol-making operations cannot begin until the application has been returned by the TTF regional director, indicating that you are in compliance. Once that happens, an Alcohol Fuel Plant Report should be filed annually with the TTF to remain in compliance.

Plant Premises and Storage

The language of legislation concerning the physical plant for alcohol production is not particularly specific. For larger commercial operations, it's simply expected that the commonsense security and protective measures taken with any business would be applied to the fuel-making business as well. With small operations, it's understood that there may not be a lot of capital investment in the plant, except for what's needed to produce alcohol. Nonetheless, a reasonable degree of protection against casual public access, vandalism and outright theft is presumed.

Many small alcohol fuel operations are conducted at whatever site is available. In an on-farm situation, there may be a vacant outbuilding or shed large enough to house the equipment. In warmer climates, a simple lean-to or pole barn with sufficient overhead clearance might be all you would need. Some operations are even conducted outdoors, though that's probably the least practical scenario unless production is seasonal, or the plant is truly experimental in nature as opposed to being set up for production.

If your operation is housed within a closed building, a lockable door may be all that's needed to satisfy the federal administrators. If it is in a lot of some type, the regulatory administrator's decision may take the site's location into account, as an urban or suburban setting would be at a greater risk for unauthorized access than a rural site might be. In this case, an enclosure such as a chain link fence may be needed to secure approval of the permit.

Alcohol storage is a factor in permitting as well. Aside from the practical aspects of tank storage — stainless versus mild steel, pump or gravity fed, venting, etc. — the administrator's concern will be that the finished product is in a locked container with hard plumbing that cannot be breached easily. Bulk storage after the fuel's been denatured (rendered unfit for consumption) can be above or below ground, and protected with a simple enclosure if you feel that an exposed tank is a little too tempting.

Here is as good a place as any to mention that fuel alcohol, even that manufactured at a small plant, is not supposed to be stored in containers of less than 5 gallons in size (except for labeled testing samples). Also, any alcohol fuel container of less than 55 gallons should be designated with a label or mark that states: Warning — Fuel Alcohol — May Be Harmful or Fatal if Swallowed.

Fig. 3.3: *The original federal Bureau of Alcohol and Tobacco pamphlet, "Distilled Spirits for Fuel Use." The information within is now available online at ttb.gov/industry_circulars/archives/1980/80-06.html.*

Denaturing Process

Before any distilled spirits can be removed from the manufacturing premises, they must be rendered unfit for beverage use with the addition of unleaded gasoline, kerosene, methyl isobutyl ketone, or other substance approved by the Bureau. There are at least eight denaturing formulas permitted by the TTB (listed under Title 27, Code of Federal Regulations, Part 21), and there is a provision in 27 CFR 21.91 for substitute materials on a request basis, pending written approval by the TTB Director. In April 2008, the TTB authorized the use of "straight run" gasoline as a denaturing agent for alcohol fuel producers because of its lower cost. Straight run, or blendstock, gasoline is defined by the TTB and is predominately pre-cracked or condensate gasoline.

The denaturing formulas are based on 100-gallon units of alcohol. As a practical matter, for most small producers, unleaded gasoline or kerosene will be the denaturant of choice, because it's easiest to come by. It's required that at least 2 gallons of denaturant be added to each 100 gallons of at least 195-proof alcohol fuel manufactured.

At one time, the Bureau published a pamphlet, ATF P-5000.5, "Distilled Spirits for Fuel Use," which summarized this and the other information that follows concerning record keeping, bonds and so forth. Those details are now available online at ecfr.gpoaccess.gov, which allows you to search through the regulations by chapter, part and section.

Essential Record Keeping

Record keeping and gauging the alcohol produced by volume are explained in the Code of Federal Regulations (CFR) referenced above. Where applicable, the rules differentiate between beverage alcohol and fuel alcohol, and fuel alcohol (once it's denatured) can be accounted for in wine gallons, so it's not technically necessary to determine the proof of the ethanol manufactured. However, you'll want to have an accurate measurement of the strength and volume of alcohol you produce, so you'll probably need to get an accurate thermometer, a hydrometer, and a laboratory-type graduated cylinder to take periodic proof readings, particularly if you store undenatured product. (The ethanol's temperature affects proof readings, so "true proof" can be established with the chart presented in Appendix A.)

Holders of an Alcohol Fuel Producer permit are required to file operational reports, but Small plant proprietors only need to prepare the report annually, within 30 days of December 31 of each calendar year. The Alcohol Fuel Plant Report F 5110.75 asks for information on proof-gallons of spirits produced, quantity and type of material used as a denaturant, and wine gallons of fuel alcohol manufactured.

Insurance and Bonding

The bonding requirement originally enforced under the Experimental Distilled Spirits Producer rules was eliminated for Small plant producers some years ago. However, if you plan to upgrade your Small plant permit to a Medium plant permit, a minimum bond of $2,000 will be required, extending to $50,000 for the maximum 500,000 proof-gallon limit imposed on Medium plants.

Insurance is not necessary under the federal AFP permit, but for those leasing or paying a mortgage on facilities, or for those who employ or use non-owners in production, a comprehensive policy and liability insurance or some type of commercial package would be highly recommended. You should keep in mind that alcohol fuel production is not inherently dangerous just because it's a fuel, especially when compared to gasoline. Neat (100 percent) ethanol's autoignition temperature is 793°F, compared to about 430°F for gasoline. Its flashpoint is also 100 degrees higher, and ethanol's latent heat of vaporization — a measure of how much energy it takes to make it evaporate — is more than twice as high. Nonetheless, the fuel factor, plus the method used to fire your boiler, may influence the coverage you're able to get with conventional policies.

Caught in the Act

After our first distillery was hastily set up right on First Avenue in downtown Hendersonville, North Carolina, *Mother Earth*'s founder, John Shuttleworth, immediately had the magazine's promotion crew organize a press conference to announce that we were turning corn into a usable fuel. The local newspapers, TV stations, and other media were on hand for a quick lesson in distillation and a demonstration of our first alcohol-powered vehicle, a 1970 Chevy pickup.

This kind of dog-and-pony show is always a little off-beat, and usually interesting because it's different — in this case, different enough to catch the eye of a producer at one of the larger network affiliates in the state, which decided to run the story in the bigger markets including the TV station in Raleigh, our state capitol.

NED DOYLE

Fig. 3.4: *A press conference held in Hendersonville, North Carolina to publicize the manufacture and use of small-scale alcohol fuel.*

Now it so happened that our little shade-tree distillery, cranking out clear, 160-proof ethanol, was directly across the way from the state Alcohol Beverage Control Commission store, where folks buy their liquor and where the local ABC supervisor maintains his office. A person standing at the threshold of that office could look right across into our open shop bay door and see anything they cared to inside.

To appreciate what occurred next, you have to understand that this was the Tarheel State of the 1970s, when many counties were dry and illicit moonshining operations were very much a part of the rural economy. While finishing his supper, the local ABC supervisor got a blistering call from his chief, the head honcho in Raleigh, who moments before had been relaxing in front of the TV news. The conversation went something like this: "These 'ol boys are making white liquor right under your nose and you don't even know it? What in the Sam Hill are you people *doin'* down there?"

The very next morning we had a visit with our local ABC supervisor who'd arranged an inspection with the state revenuers straight away. While we were opening our records and accounting for every drop of alcohol we'd made, a cluster of field officers were shaking their heads in disbelief at what was before them. "We bust things like this up every month. . . . I don't see how you could be running this setup in broad daylight!" Eventually, cooler heads prevailed, and — with tidy records and a brief review of NC General Statutes on alcohol *fuel* production — we were given a clean bill of health and told to carry on.

State and Local Jurisdiction

The federal Alcohol Fuel Producer permit does not give you authorization to operate in violation of state and local laws. In fact, state and particularly local regulations can be far stricter (and more repressive) than the enlightened federal laws, which were actually drafted for legislation in 1979 by forward-thinking staff in the BATF's regulatory cadre.

In those early days, part of the problem was that state regulators hadn't caught up with federal legislation regarding alcohol used as fuel. The mindset was still very much in the domain of beverage alcohol and its subsequent tax revenues, and it took the state bureaucracies a little time to get up to speed, both with legislation and enforcement.

Today, any questions tend to be more related to environmental issues, fire and safety codes, and insurance liability than to revenue concerns. If you keep accurate production records, adhere to the denaturing formulas, and stay completely aboveboard with your ethanol product, the state revenue department *should* have no complaints with a small fuel producer — but there's no guarantee that overzealous enforcement may not surface in some instances. Keep in mind, as well, that the bureaucratic nature of legislation may itself shine through in governments built upon that sort of thing: more regulation means greater oversight, which brings more tax revenue, which leads to more regulation, and the beat goes on.

As of this writing, all 50 states and the Canadian provinces have made provisions for legitimate alcohol fuel production, and 27 states and six provinces have incentives of one type or another in place for alcohol fuel producers. By and large, current state rules concerning compliance defer to or follow the federal regulations, often citing a specific code as a guideline.

In my state of North Carolina, a potential alcohol fuel producer must contact the state Alcohol Beverage Control Commission for notification of intent, and to get a set of guidelines for alcohol production.

Fire Code and Zoning

To be sure, compliance with local fire codes is a legitimate and necessary requirement for any manufacturing facility. Unfortunately, it can also be a tool used to wield authority when none other exists. The presence of a commercial or micro-industrial operation can raise hackles in even quasi-suburban settings. Let's face it — having neighbors in the business of manufacturing does not usually enhance residential property values. But, at the same time, all this is usually predicated upon scale. A very small operation in a backyard garage is not likely to create a problem unless someone complains about noise or odor, neither of which should be a major presence if you design your plant properly.

Before investing any real time or finances in a serious operation, it would be wise to do some research into land use and zoning regulations in the area where you intend to operate. In rural counties, restrictions should be effectively non-existent, and the impact of your operation not much different from that of a typical farm. In more urban locations, even a small fuel distillery could require permitting, and certain types of manufacturing could be prohibited. You should be prepared to face electrical and plumbing inspections as well as fire inspections.

Speaking of fire, by any standard measure, ethanol is safer than gasoline with regard to ignition. Gasoline evaporates and ignites at a much lower temperature. But two things that could raise the concerns of your fire marshal are (1) it burns with an almost invisible flame, even at lower proofs (this characteristic is altered once the fuel is denatured); and (2) the foams used on conventional gasoline fires are dissolved by alcohol. Still, plain water — which is never used on petroleum fires because it spreads the blaze — works against alcohol given enough volume to deeply dilute the fuel. Common Type ABC fire extinguishers will put out small alcohol fires as well. And special polymer foams are available to extinguish larger conflagrations.

Any reticence among fire-prevention professionals to support alcohol fuels really stems from the production of larger, industrial-sized manufacturing plants that fill tank trucks and rail cars with tens of thousands of gallons of 195-proof alcohol. Given safe storage techniques — aboveground outdoor tanks, with proper venting and sealed fittings — several hundred gallons of fuel alcohol is not a major hazard.

Richard Freudenberger

Fig. 3.5: *Smaller distillation equipment can utilize wood fuel for heat very economically.*

Environmental Impact

Item 15 in the federal AFP application asks about compliance with the federal Clean Water Act, which legislates discharge into navigable waters of the US. Likewise, state or local authorities will probably want assurances that your operation will not create an environmental hazard, and even some property lease agreements include terms to that effect.

From an economic standpoint, it would not be to an alcohol fuel producer's benefit to waste anything in discharge. I'm not speaking of fines here, either. Almost everything produced in an alcohol fuel operation can be used or recycled, from wet distiller's grains to spent mash to hot water from the still and condensers. Carbon dioxide from the fermentation process can be recovered and bottled, and "sweet water" — what's left of the beer after all the usable alcohol has been distilled out — can be fed to livestock or reused for the next batch.

How you heat the still could also impact an environmental position, but on a small scale many of the regulations simply do not apply. Natural gas is a conventional heat source, with no repercussions. Wood-fired and even alcohol-fired distillation is common in small operations, and co-generation heat has also been used with some degree of success.

FOUR

Do-It-Yourself Economics

The small-scale producer • Sourcing raw materials • Buy it or build it? • The value of your time • Calculating cost per gallon

Like any other product, ethanol is a commodity and its value is determined by what goes into it. As a small-scale fuel producer, your motivation may be more than economic — you could be aiming to reduce your carbon footprint by using a renewable fuel, or working toward becoming more self-sufficient — but like it or not, you're still subject to the same dynamics that manipulate the market for everything else we buy.

This isn't to say you have no control over the cost of your alcohol fuel product — far from it. But you'll probably have to accept certain realities and make sacrifices in some areas to balance the value of the fuel in your favor. Make no mistake about it: Producing fuel alcohol can be an inspiring and even exhilarating experience, but it is also a lot of work. Making the commitment required to set up what amounts to a private distillery is not a casual undertaking, and the time needed to collect raw materials and operate the equipment is significant.

PAINTERLAND FARMS

Fig. 4.1: *A farm-scale ethanol fuel operation requires a commitment to time and equipment. These fermentation vats work on a continuous basis.*

The economies of scale tend to bedevil everyone in planning a small operation because so many variables are entwined. How do you calculate how much fuel you'll need? Should you convert one vehicle for straight ethanol use or try an E-85 blend? If you'd rather start small, how efficient can a smaller operation be? Would it be better to go with a larger still to keep costs down? Will your feedstock and fuel sources be consistent? If not, how will that change ethanol output and costs?

None of these questions have a definitive answer because each is dependent on some other part of the operation. Fortunately, as a small player, you'll enjoy a flexibility that larger operations don't have, and you'll also be somewhat freer to take more of a "seat of the pants" approach to your effort.

Table 4.1
Sample Yields of Selected Crops

Feedstock	Lbs./Unit	Fermentable Percentage	Yield in Gal./Unit
Corn, Field	56	72	2.5
Rye	56	69	2.2
Rice	2000	77	85.0
Wheat	60	71	2.4
Barley	48	54	2.0
Sorghum	56	74	.50
Sweet Potatoes	55	24	1.0
Sugar Beets	2000	17	21.0
J. Artichoke	2000	17	25.0
Potatoes	60	16	.78
Apples	48	13	.34
Pears	56	25	.28
Grapes	2000	18	21.6
Citrus Waste	2000	7	16.0
Sugar Cane	2000	13	17.9

Note: Alcohol yields at 190 proof

Ideally, your ethanol plant would be part of a farm or market-growing venture, for two reasons. First, as a grower you'd already have a familiarity with the day-to-day practices that agriculture entails. This includes working within a routine, searching for markets, dealing with equipment in both fair and inclement weather, and quite importantly, improvising when necessary to keep things running smoothly. As anyone who has worked the land can tell you, the most successful farmers are well-rounded Renaissance people who can roll with the punches and take things in stride.

Second, a working farm provides a ready-made outlet for the manufactured fuel and its by-products. Most any internal-combustion engine or heating appliances can be adapted to run on alcohol — this inventory includes tractors, trucks, pumps, generators, burners and furnaces — and the residual material from mash production contains enough nutrient to supplement normal livestock feed.

If agriculture is not in your background, it's still possible to manufacture alcohol, even economically, provided you have a reliable source of raw material, or feedstock. As you'll see in the next chapter, there are many viable candidates for ethanol production, including both sugar and starch crops. Residues from canning and juicing operations, even far from the farm, are also distinct possibilities. Realistically, it would be difficult to carry on much more than an experimental venture in a confined space such a suburban backyard, but it's still possible. Ideally, a rural setting or a location where there's room to expand and function without interference would be the better choice.

Sourcing Raw Materials

Finding a reliable and consistent source for feedstock material can be a real challenge. Chapter 5 will address the distinction between sugar crops such as cane, sugar beets and fruit juices and starch-based crops such as corn, sorghum, grains and potatoes. For now, it's enough to say that certain plants produce more starch or sugar per ton or per acre than others, and given the right cost, crops with more concentrated nutrients are the best choice.

To complicate matters, though, is the fact that the equipment needed to process the raw material varies by crop. Grain-grinding machinery is quite a bit different from the extractive equipment used to process sugar beets. Unless you can cultivate a reliable source of feedstock, it would be unwise to invest in any specific equipment. Consider, instead, renting (or leasing) that equipment if possible, or look into using the services of a local co-op.

If you live in a rural community where processing and packing houses exist, you may find that reclaiming surplus and spoilage from these operations makes the best economic sense. Approached properly, most cooperatives and private processing facilities should be willing to negotiate an attractive arrangement — a deal, if you will — that would allow you to test the value of their spoilage as a feedstock, subject to performance results over a specific period of time.

Too, you can always try making arrangements with individual farmers, perhaps in exchange for culling waste from fields and orchards, which will provide you with the needed feedstock material, at least on a temporary basis while you establish its feasibility.

Storage can be an issue with certain crops. Some products should be processed within a few months of harvest, and if they are not, they need to be dried sufficiently to store. Drying and storage come at additional cost, and are best both avoided. You will, of course, have to make some provision for containing your feedstock on a day-to-day basis to keep the operation running smoothly, especially if you plan to operate the still in batches rather than on a continuous basis.

Buy It or Build It?

For the small-scale fuel producer, many still designs are so basic that it's much simpler and far less expensive to build the equipment rather than to buy it. This is especially true of small-capacity operations. Costly stainless steel components aren't needed at this scale — ordinary mild steel pipe will do for the columns and water lines, and in some applications plastic piping can be used. Likewise, tanks and vats need not be anything special, but for those elements, it's often cheaper to just buy used equipment at a farm auction (stainless steel dairy storage and processor tanks are common auction items).

If you have welding skills and a place to work, you're way ahead of the game. For the kind of components involved, there's no real reason to use new materials. Any salvage or metal scrap yard is likely to produce the sort of parts you'll need. If you're not fussy, an old oil tank can make a decent boiler vat, and similar liquid storage containers can be adapted to serve as agitated mash cookers. Many components are make-do items from other applications, so you'll have to use a creative eye when shopping for good candidates. Unfortunately, many manufactured steel items — particularly stainless steel — have increased in value in the pre-owned marketplace because there is an increased foreign market for quality steel salvage in general and for well-made American products in particular, especially among developing nations. The plumbing parts are for the most part standard off-the-shelf items.

Fig. 4.2: *Water tanks harvested from a scrapyard for repurposing as fermentation cookers and storage vats in a backyard operation.*

CLARENCE GOOSEN

Paying for the services of a professional welder will increase the cost of the equipment considerably and perhaps even double it. You can trim expenses by locating all the materials yourself and preparing the parts to be fit and welded prior to delivering the job. The less the welder has to do in shaping, fitting and grinding, the less time he or she will spend on the project, reducing the hourly charge. This prep work is not a particularly high-skilled endeavor, and the investment in tools is very reasonable at this stage, so you might consider taking this approach and saving a few dollars in the bargain.

The Value of your Time

Unfortunately for some of us, we are blessed with a desire to learn and accomplish rather than driven to make a profit. Such is the case for those working at the preliminary stages of setting up a home-scale distillery. Still, putting a lot of sweat equity into your ethanol project is a sound decision, especially for those who aren't fully committed to the idea of making large volumes of fuel alcohol. It reduces the amount of monetary investment involved (and thus the risk) and also provides you an intimate familiarity with the equipment that you'd never experience simply by purchasing it.

Once you're at the point of producing ethanol, you should place some value on your time, even if it is minimal. Assigning a cost per hour to your labor in collecting and processing feedstock, maintaining the distillery's operation, and handling the ethanol product and its record keeping will allow you to honestly and accurately calculate what it costs to be independent of the normal petroleum fuel network.

Calculating Cost per Gallon

It is not that difficult to figure out what it will cost you to make a gallon of ethanol fuel, given some degree of stability in the cost of your cooking/heating fuel and feedstock sources. In a traditional farming operation, the costs of production are well established and independent of yield per acre and market value of the crop, until it comes time to calculate the actual level of profit.[1]

The situation is similar with ethanol fuel, though many producers, particularly those working with spoilage and processing surplus, will not be concerned with crop yields other than their value in starch or sugars.

In order to keep your computation consistent, it is prudent to convert your alcohol yield to a standard proof measure, especially if you're drawing varying proof percentages of ethanol from your still. I established in Chapter 3 how the revenue authorities calculate ethanol measure for the purposes of taxation; you should use a similar method to determine the value of your fuel.

For example, if you've made 100 gallons of 185-proof ethanol in one run and 50 gallons of 190-proof ethanol in another, you can conclude that your yield is 140 gallons of 100 percent ethanol. The actual product, of course, is not that pure, but you're simply establishing a standard common denominator you can work with for the purposes of calculation. Table 4.2 illustrates how to make these calculations.

Once that's established, you can determine the cost of your raw material feedstock, calculate the cost of transporting it to your work site, and subtract the value of any by-product yield, whether it's sold as distiller's grain or used for yourself at fair market value. This would include carbon dioxide for bottling, and any cellulosic co-products, which can be fermented to produce methane gas or dried for boiler fuel.[2]

At this point you have a net feedstock value, for which you must now factor the cost of conversion to ethanol. The operating expenditures involved in this process include the cost of supplies such as enzymes and yeast, the cost of fuel to cook the mash and heat the distillation boiler, and the cost of insurance, licensing and any financing. These are added to the net feedstock figure to give you the cost of ethanol prior to adjustments for depreciation and other miscellaneous costs such as electricity for pumping, maintenance and repairs. Depreciation may be the cost of any leased equipment or machinery purchased, which can be extended or amortized over a given period, generally five years. Labor costs can also be considered here, though they may change with increased or decreased production.

Table 4.2
Proof Measure Calculation Formula

Number of gallons X Proof strength of product ÷ 100 = Proof-Gallon measure.

Example: Batch One is 50 gallons at 190 proof.

50 X 190 = 9,500 ÷ 100 = 95 Proof Gallons

To establish a common denominator to evaluate the cost of a series of batches, the runs can be averaged:

Example: Batch One is 100 gallons at 185 proof. Batch Two is 50 gallons at 190 proof

100 X 185 = 18,500	50 X 190 = 9,500
18,500 ÷ 100 = 185 Proof-Gallons	9,500 ÷ 100 = 95 Proof-Gallons

185 + 95 = 280 ÷ 2 = 140 gallons at 100 Proof

If you prefer to work in percentages, the same result can be obtained using the identical example:

185 proof = 92.5 percent	190 proof = 95 percent
.925 X 100 gallons = 92.5 gallons	.95 X 50 gallons = 47.5 gallons

Total = 140 gallons at 100 Proof

The total fuel cost is then established by adding the adjusted costs above to the pre-adjusted cost of your ethanol to get a net cost.

Table 4.3
Calculating Cost per Gallon

Cost/ Raw feedstock	$137.48
Cost/ Transportation	5.44
Subtotal	$142.92
Credit/ Byproducts	34.80
Net Feedstock Cost	$108.12
Cost/ Enzymes and yeast	5.23
Cost/ Heat source fuel	2.14
Costs/ Fixed	1.66
Pre-adjusted Total	$117.15
Cost/ Depreciation and labor	8.65
Net Cost	$135.80
Yield/ Gallons pure alcohol	76.2
Cost of Alcohol Fuel	$ 1.78

Note: Figures are representative. Fuel, labor and feedstock costs can vary widely depending upon particular circumstances.

Dividing this figure by the number of pure ethanol gallons (not actual gallons) will give you the cost per gallon of your hard-earned product.

Beginning in 2005, an enhanced Small Producer Tax Credit became available with passage of the Energy Policy Act of 2005. Section 40 of the US Internal Revenue Code now allows an eligible small ethanol producer, defined as one manufacturing less than 60 million gallons per year, a federal income tax credit equal to $ .10 per gallon for the first 15 million gallons produced. Individual states may have other such incentives for small producers as well.

In the following chapter, we'll take a close look at what is arguably the most significant factor in establishing both the economic and fundamental value of your alcohol fuel — the feedstock and raw materials.

FIVE

Feedstocks and Raw Materials

Starches, sugars and cellulose • Sugar crops • Starch crops • Residues • Waste products • The dynamics of yield

Feedstocks are the raw materials used to make ethanol. In theory at least, most plants and agricultural products can be used as a feedstock. Some crops readily yield the simple sugars needed to make alcohol; others are starches and must be broken down from their complex form to produce those sugars. Certain crops have a high yield per acre but may require special harvesting equipment or have other adverse demands that diminish their appeal as an alcohol-producing feedstock. Usable crops suitable for forage can generally be grown on marginal land, and still others are suitable for ethanol production but are otherwise removed from mainstream agriculture.

Still, the basic ethanol production process is similar for all fermentable materials. Fermentation occurs when microorganisms (in this case, yeasts) convert simple sugars in a liquid mash mixture to ethyl alcohol and carbon dioxide, in addition to creating some heat and an enzyme called adenosine triphosphate. The fermented liquid, called "beer," contains a low percentage of alcohol by volume, and therefore must go through a distillation process to increase the percentage needed to produce a high-proof fuel-grade ethanol.

Yeast is a single-cell microorganism, more precisely a *Saccharomyces* fungus, that produces alcohol and carbon dioxide in anaerobic conditions. Yeast does its work of transforming carbohydrates in the feedstock to alcohol once those starches and sugars are in a form the yeast can use. It is enzymes — catalytic proteins that promote the chemical processes needed to break down starches and the more complex sugars to usable form — that make that happen. Generally, two types of enzymes

Table 5.1
Average Alcohol Yield per Weight from Raw Material

Feedstock Material	Yield in Gal./Cwt.	Yield in Gal./Ton
Wheat	4.25	85.0
Corn, Field	4.20	84.0
Buckwheat	4.17	83.4
Raisins	4.07	81.4
Sorghum grain	3.97	79.5
Rice	3.97	79.5
Barley	3.96	79.2
Dates, dry	3.95	79.0
Rye	3.93	78.6
Prunes, dry	3.60	72.0
Molasses, blackstrap	3.52	70.4
Sorghum cane	3.52	70.4
Oats	3.18	63.6
Figs, dry	2.95	59.0
Soybeans	2.44	48.8
Sweet Potatoes	1.71	34.2
Crabapples	1.29	25.8
Yams	1.36	27.3
Peanuts	1.35	27.0
Potatoes	1.14	22.9
Sugar Beets	1.10	22.1
Figs, fresh	1.05	21.0
J. Artichoke	1.00	20.0
Citrus Waste	0.83	16.6
Pineapples	0.78	15.6
Cranberries	0.78	15.6
Sugar Cane	0.76	15.2
Grapes	0.75	15.1
Apples	0.72	14.4
Apricots	0.68	13.6
Pears	0.57	11.5
Peaches	0.57	11.5
Plums	0.54	10.9
Pumpkins	0.49	9.8
Carrots	0.49	9.8
Whey	0.37	7.4

Note: Alcohol yields at 190 proof. Yields complied from *Goosen's EtOH Fuel Book (1980)* and USDA sources. Short Hundredweight (Cwt.) equals 100 pounds US. Short Ton equals 2,000 pounds US.

Table 5.2
Average Alcohol Yield Per Acre from Raw Material

Feedstock Material	Yield in Gal./Acre
Sugar Beets	287 - 412
Sugar Cane	268 - 555
Corn, Field	214 - 390
J. Artichoke	180 - 613
Potatoes	178 - 299
Sweet Potatoes	141 - 190
Apples	140
Dates, dry	126
Carrots	121
Raisins	102
Yams	94
Grapes	90
Peaches	84
Prunes, dry	83
Pineapples	78
Pumpkins	78
Cranberries	70
Rice	66 - 175
Soybeans	58
Pears	49
Barley	48 - 83
Molasses, blackstrap	45
Apricots	41
Peanuts	40
Oats	36- 57
Sorghum grain	35 - 125
Buckwheat	34
Wheat	33 - 79
Figs, fresh	31
Figs, dry	29
Sorghum cane	26 - 500*
Rye	24 - 54
Plums	22
Citrus Waste	N/A
Whey	N/A
Crabapples	N/A

Note: Alcohol yields at 199.5 proof. Figures rounded to nearest whole number. Yields based on multiple sources including USDA Misc. Pub. 327, December 1938 and USDA Agricultural Statistics, 1978 and 2003.
* Lipinski, E.S., "Fuels From Sugar Crops." 1979

are used, an alpha amylase to reduce the starches initially, and a glucoamylase to complete the conversion of starches to sugar. Chapter 6 explains this complex relationship in more detail.

The simplest and most common sugar found in plant materials is glucose, a compound containing six carbons, twelve hydrogens and six oxygens, expressed as $C_6H_{12}O_6$. This simple sugar is a monosaccharide, or fermentable sugar, and all agricultural crops and crop residues contain these sugars or compounds of them.

Types of Feedstocks

Depending upon the composition, feedstocks are classified into three different categories: sugar crops, starch crops and cellulosic crops such as forage and residues. With today's technology and the rising cost of agricultural products and energy, the traditional models of ethanol production, i.e., using corn-derived starch as a feedstock and natural gas as boiler fuel, must be re-evaluated with a critical eye. A more wholistic approach can be both economical

RICHARD FREUDENBERGER

Fig. 5.2: *A glucose molecule.*

Embden-Meyerhof Pathway

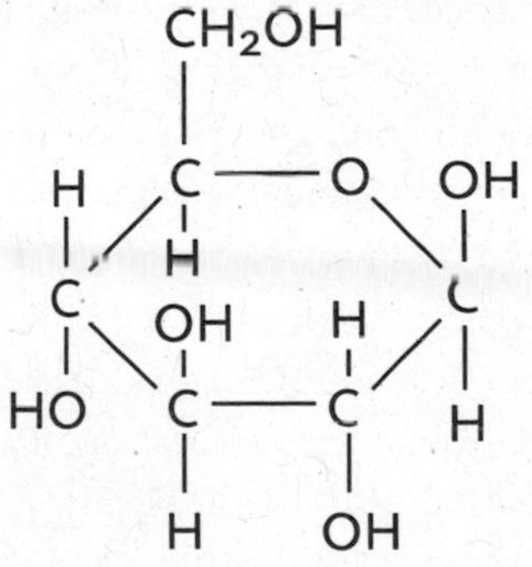

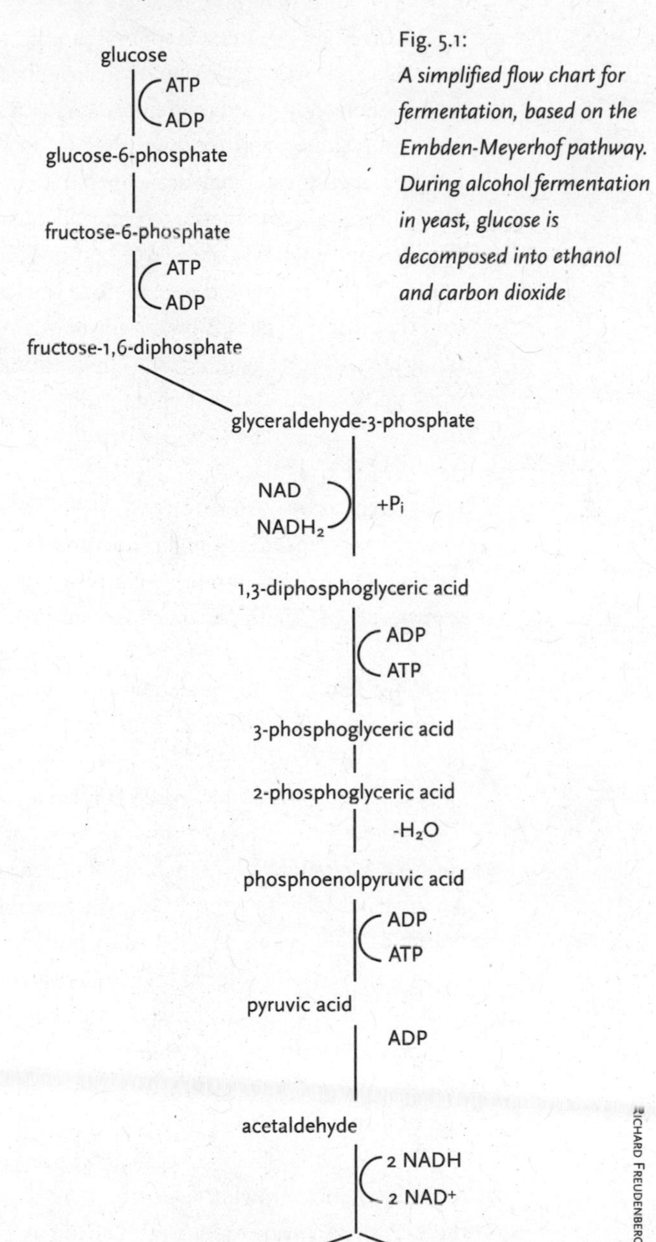

Fig. 5.1: *A simplified flow chart for fermentation, based on the Embden-Meyerhof pathway. During alcohol fermentation in yeast, glucose is decomposed into ethanol and carbon dioxide*

RICHARD FREUDENBERGER

and sustainable — and a whole lot healthier for the environment.

Accordingly, progressive ethanol producers in the US and especially in developing countries are evaluating non-traditional crops, agricultural cull and residues, and food industry waste as potential feedstocks. Small-scale operations are especially receptive to these alternatives because they are more flexible than large commercial or industrial interests and have more to gain by experimenting. What follows is a broad sampling of conventional and non-conventional stock with potential for ethanol production.

Sugar Crops

Most sugar crops are readily identifiable by their names. Sugar beets, sugar cane, and sweet sorghum are some common examples. Other crops, such as Jerusalem artichokes and fruit crops (apples, peaches and pears to name a few), are also high in fermentable monosaccharides, and are feasible candidates for ethanol production.

The six-carbon simple sugars in the majority of these crops occur individually or are connected in pairs. This non-complex structure readily lends itself to fermentation because the feedstock only has to be crushed or milled to release the sugar in a form that the yeast can work with. The equipment used in preparation is commonly available and widespread in agriculture, and includes grinders, hammermills and extraction machinery.

The ease with which these sugars convert is something of a drawback, however, because these crops are prone to spoilage. The microorganisms responsible for spoilage flourish on sugars and the high moisture content of sugar crops, and in the presence of air will eventually convert the sugars to acetic acid and carbon dioxide in storage.

That problem can be addressed either by using the feedstock directly after harvest, or by drying it through evaporation to remove the moisture so it can be stored safely. From an economic standpoint, drying is energy intensive and incurs additional costs in equipment. Solar dehydrators can reduce operating costs over conventional evaporators, but still require capital investment.

SUGAR BEETS

Sugar beets tolerate a wide range of soil and climatic conditions and are widely cultivated. They are especially suited to cooler climates where other crops may not fare as well. Processing requirements are minimal, with mechanical milling followed by pressing being the process of choice for extracting the sugary juice. Slicing is also an option. The beets contain about 16 percent sugar and will ferment at a temperature of 85°F in a short time — between 8 and 24 hours or so. Beets can also be hammermilled whole and mixed with water to form a slurry, but will then need to be cooked for at least an hour before introducing yeast. As a root crop it requires rotation with non-root cultivars to discourage parasitic nematodes. Ethanol yield is relatively high, at up to 750 gallons per acre in some cases.[1] Yields of 22 gallons per ton of product are common.

Sugar beets also have the benefit of storing well in piles up to four months time, and providing valuable co-products. The post-processing residual pulp can be used as distiller's feed, and the beet tops provide organic material for fertilizer.

Sugar Cane

In the US, this crop is generally limited to four states (Texas, Florida, Louisiana and Hawaii) because of climatic requirements, but in countries such as Brazil, the ethanol industry has been built upon it. It offers among the highest yields of ethanol per acre — as much as ten times that of corn under the right conditions[2] — and also provides a significant yield of residual biomass (bagasse), which can be used for boiler fuel and co-generation after mechanical or manual processing. In addition, the nutrient-rich stillage left over from distillation is used as liquid fertilizer for the next season's crop after the cane has been harvested.

Sugar cane is a perennial; it is planted once and then harvested for five or ten seasons afterward. Much is made of sugar cane for its alcohol potential, and subsequently it is held as a shining example for the ethanol economy. Realistically, though, cane cannot thrive at temperatures below 45°F, and will succumb at temperatures well above freezing, so its potential as a small-scale ethanol crop in most of the US and Canada is limited. Although certain varieties of the crop can be planted and harvested annually, they require a long growing season between the end of spring frost and the onset of late frost. Though this expands the potential acreage for cultivation, it is also a more labor-intensive approach.

In Brazil, the traditional harvest of sugar cane is environmentally inappropriate in that the field crop is first burned to remove the foliage and expose the stalk, or ratoon, of the plant. Following that, an army of laborers manually cuts and stacks the stalks for processing. In more recent times, workers have been employed in removing the leaves (which contain useable biomass) by hand, a practice that provides jobs and reduces particulate pollution. On well-capitalized plantations, harvesting equipment is used to gather the cane and leaves, a time-efficient but costly option.

The point is that either labor or capital investment is required to propagate sugar cane, as is the case with any other crop. But yields from the cane juice regularly run around 750 gallons per acre, and the bagasse and leaves — comprising well over half the plant's weight — have value as fuel and as cellulosic-conversion feedstock.

Sugar cane is typically crushed in a drum crusher or roller mill, which squeezes out and collects the juice and sends the stalks out a conveyor chute for drying. Crushers are manufactured on every scale and can occasionally be found as salvage. Another method of extracting the sugar is known as diffusion, in which thin sections of cane are steeped and heated in a solution to draw the sugars from the substrate material into the solution. The enriched solution is passed through several stages, drawing additional sugar from a fresh substrate supply at each stage. When the initial supply is exhausted of its sugar, it is removed for use as bagasse or livestock feed and fresh cane sections are added. New substrate is introduced on this rotating basis until the cane supply is exhausted.

Sugar cane only stores for a couple of months in stalk form, and should be used immediately once juiced. Fermentation is rapid, and will be complete in less than 24 hours. Cooking is not required.

SWEET SORGHUM *(Sorghum bicolor)*

Two kinds of sorghum are grown in North America: a grain sorghum known as sorghum bicolor and a stalk-like cane crop known as sorgo or sweet sorghum. Either a cane crop or a variety cultivated for grain is a viable candidate for ethanol production in temperate regions. A good yield of ethanol per acre and a generous tolerance for varied climatic and soil environments give sweet sorghum a welcome flexibility as a feedstock. Some varieties yield both grain (milo) and sugar, which further increases its value as a feedstock.

In processing, the plant can be harvested and stored whole, or processed to remove the sugar juices upon harvest. It can also be mechanically chopped, and the green fodder stored as silage. The high protein content of its leaves and residual fiber make it an excellent livestock feed. Moreover, like sugar cane, the residual matter from sweet sorghum can be burned as a heat source for ethanol processing.

Sorghum has high sugar content and can produce two crop cycles per season in warmer climates. Alcohol yield can be between 400 and 600 gallons per acre.[3] Extracting juice from the cane can be accomplished by crushing short sections in a roller mill, but the fiber must be separated. On a small scale, shredding and fermenting the product whole may be more cost effective.

Grain (milo) sorghum may offer a significant yield in some varieties. After harvesting, the grain should be ground and a liquid mash prepared, then heated and hydrolyzed, or broken down, as with other grains. The section on starch crops below describes this process in more detail.

MOLASSES

Molasses is not a crop in itself, but is an extraction from sugar cane or beet juice, and is what's left over after crystals are removed from the juice to make table and culinary sugar. Molasses is used a supplementary cattle feed, and prior to the Second World War was one of the primary sources of ethanol manufacture for commercial alcohol.[4]

Blackstrap molasses, the concentrated by-product of sugar crystal production, is nearly ready-made for ethanol production. The sugar content of molasses will vary according to its source, but ranges from 35 to 50 percent. The thick syrup can be a potential feedstock if it's readily available, but costs can be high and tend to fluctuate on the open market. You can get an idea of whether molasses will be an economically feasible feedstock by looking at its ethanol yield — 70 gallons of 95 percent pure alcohol per ton.[5] At a market price of say $100 a ton, it probably isn't feasible — but when costs go down to the $50 level, there may be enough leeway to consider it on a small scale.

Molasses requires heating and hot water dilution to make the mash solution. Pumping unheated molasses can be a challenge without pre-warming or the proper pump. The mash, once diluted to 15 percent sugar level, may require the addition of some yeast food (more on that in Chapter 6). After the initial inoculation of yeast, fermentation should occur completely within 48 hours.

Jerusalem Artichokes

The Jerusalem artichoke is neither from Palestine nor is it an artichoke, but a member of the sunflower family that is well adapted to the cooler climates commonly found in North America. Sometimes marketed as a sunchoke, it thrives in a variety of soils and adapts well to marginal land. The Jerusalem artichoke is a tuber that manufactures its sugar in top growth and stores it in the tuber and roots. Most of the carbohydrates in the tuber are in the form of inulin, a non-digestible polysaccharide. It's also a perennial, so tubers left in the field can produce the following season's crop without seeding. It has no serious insect pests, and Sclerotinia wilt is the only disease known to affect yield.

Because this cultivar has historically been a tuber crop, equipment and technique has developed around that use. However, for the most part, the sugar produced by the leaves does not migrate to the tuber until late in the plant's life cycle. By harvesting the stalk as one would corn or sweet sorghum, the tuber can be left in the ground for future production, while sugar is gleaned from the stalk through a conventional milling process. This practice does away with the bulk of tillage and soil erosion associated with annual crops. Jerusalem artichokes store well in cold ground, and can be stored in high-humidity conditions at lower temperatures above freezing. The tubers can be sliced into thin discs and sun-dried to store indefinitely for future sugar diffusion (see Chapter 6), or as livestock feed. The dried leaf tops can be stored like any dried forage.

It is also possible to hydrolyze shredded tubers using heat and sulfuric acid, but a commercial inulinase enzyme that breaks down the fructose and other sugars in inulin is now available as well. (See Appendix B for resources.) Conventional yeasts can then be used to convert the normal glucose and fructose sugars to ethanol. The shredded substrate is briefly boiled in water with the addition of some inulinase enzyme. The mash is then cooled to 150°F, the recommended remaining dosage of inulinase added, and the mixture maintained at that temperature for 12 hours.

There are also specialized yeasts — *Kluyveromyces marxianus* and *Kluyveromyces fragilis* are just two examples — that ferment inulin directly to ethanol. These specialized yeasts are not commonly available, but are cultured for use with cheese whey (see below) to supplement water in the fermenting process. The whey contains lactose, which provides additional sugars digestible by the yeasts.

Both the tops and tubers of Jerusalem artichokes are promising candidates for ethanol production even on a small scale, and they offer excellent alcohol yields. A reasonable expectation is between 14 and 20 tons per acre for tubers; at a modest yield of 21.4 gallons per ton,[6] this would amount to at least 300 gallons per acre. Higher yields (up to 600 gallons per acre) can be expected with sophisticated conversion technology.[7] The non-fermentable residue left after fermentation is rich in protein and can be used as supplementary animal feed.

FRUIT

Fruit crops must generally be evaluated on an individual basis because of their potential market value for products other than alcohol. But, as far as sugar content is concerned, most fruits are well suited for conversion. Apples, peaches, pears, apricots, figs, dates, small fruits and grapes all contain highly fermentable sugars. Spoilage and the co-products of the fruit-processing industry are ideal candidates for ethanol feedstock because the alcohol-manufacturing process makes use of waste that would otherwise have to be dealt with in a costly and environmentally sound manner.

Fermented fruit has a high nutritional content and makes a good fertilizer for gardens and orchards. The leftover mash pulp can be dewatered in a press and utilized as animal feed.

FODDER BEETS

The fodder beet is a high-yield sugar forage crop developed in New Zealand by crossing sugar beets with mangolds. Agriculturally, it is similar to the sugar beet, but it has a greater yield of fermentable sugar on a per-acre basis, and it stores well in piles (tops removed so they don't decompose), with less sugar loss than the sugar beet. Like the sugar beet, the fodder beet stores most of its sugar in the roots. This crop does not do well in marginal soil and is somewhat water intensive, but is tolerant of cold climates.

Ethanol yield varies considerably by variety, though some reports indicate production levels of 850 gallons per acre. Processing for alcohol production is similar to that of sugar beets. The fermented beet pulp is a desirable animal feed supplement with a high protein content and is an excellent replacement for corn-based Dried Distiller's Grains with Solubles (DDGS).

Starch Crops

Starch crops are more difficult to process than sugar crops because their sugar units are connected in long, branched chains. These complex starch chains must be broken down into single (or pairs of) simple six-carbon units in order for the yeast to convert the sugars to ethanol.

Fortunately, the starch conversion process is a fairly straightforward one involving the use of heat and enzymes or acids. Enzymatic hydrolysis, discussed in Chapter 6, is the process by which enzymes and water break down the starch once it's been made accessible by grinding or milling.

The chief shortcoming of using starch crops for ethanol production is the additional energy, equipment, and time or labor involved in accomplishing this breakdown process. Energy, especially, is an economic consideration that requires careful analysis because of its volatile and unpredictable costs. Small-scale producers, however, have more flexibility than large plant operators, and can work more readily with alternative heat sources such as wood, crop residues, and even solar thermal energy to meet their needs. Furthermore, advances in enzyme research have created

Ethanol from Orchard Waste

One North Central Missouri fruit farmer looked to his orchard waste to produce enough ethanol to run a significant portion of his 10-acre farm's daily operations. This includes a small Ford 9N tractor, a small farm truck, and a variety of gasoline-powered mowers and trimming equipment. Daniel West, with help from a USDA Sustainable Agriculture Research and Education grant, built upon a project that he had initiated in 2002 to dispose of excess apricots after harvest. The Wests' 1,300-tree orchard is primarily in apples, but the family also grows apricots, peaches, pears, plums and nectarines, all of which generate surplus, waste fruit.

Fig. 5.3: *Dan West's 6-inch packed-column still uses fruit waste and electric heating elements.*

After some simple distillation trials to familiarize himself with the alcohol-production process, he built a small prototype, using a fractionating column made from 3-inch copper pipe filled with glass-marble packing. The 15-gallon boiler pot was heated with a single 4,500-watt electric water heater element, and produced 1.5 gallons of 190-proof (hydrated) ethanol for each 75-minute batch run, or the equivalent of 1.2 gallons per hour.

Dan West then constructed a second-generation design using a new 500-gallon "factory second" propane tank which he was able to buy for $250. This version uses a packed stripping column as well as a 6-inch fractionating column packed with 3/8-inch ceramic marbles. The two columns work together to separate ethanol from the water-ethanol mix more efficiently than a single fractionating column is able to. (I'll explain more about distillation in Chapter 7.) The boiler vat in this larger still is heated by four electric elements, and the tank and columns are insulated during operation to control heat losses.

Economies of scale and a more effective design allowed West to produce 10 ounces of hydrated ethanol per minute, or just over 4.5 gallons per hour. Theoretically, a column of that size should produce 6 gallons per hour, and the builder sees a potential for up to 8 gph with some tweaking of the column's cooling coils and condenser.

Direct expenses related to building the still came in at $1,900, which includes $500 of labor cost. Operating costs come to $1.60 per hour for the electricity and approximately $0.20 per gallon of ethanol for the water used in the cooling condensers. The feedstock is available on site at no cost. Storage and labor is included as part of the farm's operating routine, and the only ongoing expenses are for yeasts and other supplemental supplies. West estimates the lowest cost of his product at $0.65 per gallon.

Dan West views the project as an excellent opportunity for outreach in his farm community and as a means to reduce his dependence on imported petroleum and support the use of an environmentally friendly fuel within his family's orchard operation.

Fig. 5.4: *Rye and other grains are good sources of feedstock for farm-based ethanol production.*

strains that work well at temperatures in the 85°F to 104°F range rather than the 185°F to 219°F temperatures common in a conventional process.

Starch does have an advantage that the sugar crops do not possess, and that is in storage potential. The very factors that make initial conversion difficult keep the starches from fermenting readily in storage. Once the grains are dried, it's very difficult for microorganisms to work upon the starch and proteins resident within the crop. Long-term storage is a clear benefit to ethanol production because it provides some flexibility, and opens markets to additional feedstocks.

GRAINS

This category of starch crops includes corn, wheat, barley, rye, oats, grain sorghum, and even rice and millet. Though very little has been done within the fuel alcohol field with barley, rye, oats, rice and millet, these grain crops all have a relatively high alcohol yield on a per-ton basis. Processing costs, availability, or their value as food have eliminated them from the fuel market. Conversely, corn in particular, wheat, and milo or other sorghum grains have all received some attention as ethanol fuel feedstocks. I'll profile a few below:

Corn

Corn or maize, is far and away the most abundant of the grains and has a very extensive range in North America. Despite the reputation for soil erosion that corn has developed over the years, modern management practices have reduced its impact on the soil to make it more environmentally acceptable. Like many grain crops, corn has an extremely high carbohydrate content (72 percent) and therefore an excellent alcohol yield on a weight basis; the average is something on the order of 84 gallons per ton. The per-acre yield can be 330 or more gallons for each acre harvested.

For the ethanol farmer, distiller's grains also represent a significant portion of the crop's value. Because of the high protein contained in Dried Distiller's Grains and Solubles, it makes a highly sought-after livestock feed, and this even *after* the alcohol has been extracted from the starch in the corn. Put another way, the food value of the crop is enhanced even while producing a usable liquid fuel, biomass fuel, organic matter for fertilizer, and carbon dioxide for industrial use. What most people probably don't realize is that less than ten percent of the US corn crop is used for human food. The vast majority is used domestically (or exported) for animal feed and high-fructose corn sweetener, and the alcohol comes from that.

As a starch, corn needs to be ground, cooked and broken down with water and enzymes to convert it to glucose in the hydrolysis process. Chapter 6 describes this in further detail.

Wheat

Some wheat varieties — white and soft red winter — are better for alcohol production than traditional durum and hard red spring strains, which have a higher protein and gluten content. The modern wheats are high in starch but much lower in protein. If you are not that concerned with the value of distiller's grains as feed, the high-starch varieties are less costly and may be the better choice.

Wheat will yield 85 gallons of alcohol per ton, which is about 33 bushels. Harvest yield is in the range of one ton per acre or more. Wheat can be stored in bins for up to 12 months once it's dried. Ground wheat and alpha enzymes are mixed at the start with water and heated to a temperature between 155°F and 190°F for 30 to 45 minutes, depending upon the chosen enzyme's tolerance for heat. Once the temperature is brought down to 145°F, conversion enzymes are added and the mash is maintained for 30 to 120 minutes before temperature is reduced to an acceptable range for the yeast, which is then introduced.

Table 5.3
Percentage of Sugar and Starch in Selected Grains

Grain	Percent Sugar	Percent Starch
Barley	2.5	64.6
Corn	1.8	72.0
Oats	1.6	44.5
Rye	4.5	64.0
Sorghum grain	1.4	70.2
Wheat	--	63.8

Wheat mash has a tendency to foam, a problem that's exacerbated when using the higher-protein varieties. A simple solution is to add ground corn to the wheat in a ratio of 40 percent wheat, 60 percent corn. The mash can then be processed as corn, with minimal foaming. If straight wheat is mashed, a commercial de-foaming agent will resolve the problem.

Milo

I spoke a bit about grain sorghum earlier, which in many areas is used primarily as animal feed. It's another protein crop with high starch content, yielding nearly 80 gallons of alcohol per ton. There is little difference in the mashing process between milo (and barley for that matter) and corn, perhaps with the exception of a shorter boiling period.

Tubers

Potatoes, sweet potatoes, yams and other root crops such as the tropical cassava all have enough starch content for successful alcohol production. The issue with some of these crops, however, is that their value as food makes them uneconomical in some markets except as culls. Let's look a few here:

Potatoes

Depending upon the length of storage, potatoes are 70 to 80 percent water, and contain up to 18 percent starch, with only 2 percent protein. Under controlled conditions, potatoes will keep in storage for 12 months. Typical alcohol yields are 23 gallons per ton, which calculates to 178 gallons per acre given a modest harvest. The starch granules in a potato are large and not difficult to break down, but the raw material must be ground or chopped to pulverize the bulk. Considerably less water is needed in the mashing process than with other starches due to the water already in the tuber.

There is a complete potato feedstock recipe in the following chapter, so I'll limit the information here to the fact that potato mash must be heated and hydrolyzed by enzymes. The solids in the mash remain in suspension, which can be problematic for the thorough distribution of yeast. Because of this, you may prefer to separate the solids and ferment only the liquid to encourage a more complete fermentation.

Fig. 5.5: *Sweet potatoes have a high carbohydrate content and store well once dried.*

NORTH CAROLINA SWEETPOTATO COMMISSION

Sweet Potatoes

An often overlooked cultivar and feedstock is the sweet potato, which at one time was used for alcohol fuel in southern regions. Sweet potatoes will store for months when chipped and dried, but contain 70 percent water and less than 2 percent protein when fresh. They have a greater carbohydrate content than conventional potatoes, averaging about 26 percent. This, plus the fact that there has always been inherent waste in harvesting makes them tempting as an ethanol feedstock. Alcohol yields can be up to 34 gallons per ton.

Sweet potatoes are prepared by grinding or chipping, like potatoes. It is a hydrated crop, so should not require more than ten gallons of water per hundredweight, but dried sweet potatoes may require 30 or more gallons per bushel (about 55 pounds). As a starch, they must be cooked and hydrolyzed with enzymes.

Cassava

Like sugar cane, cassava is really a tropical crop, but I'm mentioning it here because it can be grown in the Gulf States and has potential in many underdeveloped tropical countries as a fuel source. It's the turnip-like manioc root from which we extract starch for bread and tapioca, and fiber for industry. Cassava is about 35 percent starch, with a moisture content of 55 percent. Both the tops and root can be used as animal feed once dried, but the tops contain a higher degree of protein and minerals.

Cassava is a moderately water-intensive plant, but will grow in very marginal soils with minimal fertilizer. Yields on a per-acre basis depend heavily on climate and how well the crop is cultivated and irrigated, but range between 13 and 27 tons per acre for an established crop. Preparation requires peeling the root first, then coarse shredding followed by a second, finer shredding. The mash is cooked and hydrolyzed much like corn would be.

Cellulosic Crops

The food versus fuel debate has intensified research into the use of cellulosic material as a feedstock for ethanol. Utilizing cellulose rather than starch or sugar crops has the obvious appeal of allocating food to human consumption rather than earmarking it for energy use. At the same time, it encourages the cultivation of marginal lands and plants that aren't as demanding of water and fertilizers, saving those resources for actual food production.

The stalks and leaves of typical starch and sugar crops are for the most part composed of cellulose. Cellulose does contain sugar, but the individual sugar units are tied together in long chains by chemical bonds much stronger than those found in starch. Since the cellulose must be broken down into simple sugar units for the yeast to do its work, the conversion is problematic because of the additional steps and cost involved. The cellulose is enveloped by lignin, a complex compound that is part of the cell walls, which gives wood its characteristic stiffness and strength.

Lignin is very resistant to hydrolysis, both enzymatic and by acids. Ongoing research into cellulosic conversion by biotech firms such as Canada's Iogen (see Chapter 10), and universities such as Perdue and Penn State, has produced demonstration facilities that successfully utilize cutting-edge technology, but the cost of converting liquefied cellulose into usable, fermentable sugars is still well beyond the means and scope of a small-scale production facility. Of course, as the cost of conventional petroleum fuels increases, cellulose conversion and its associated enzyme development becomes more attractive, and eventually it will be economically viable.

To give you an idea of how tough a nut this has been to crack, research done in the late 1970s and subsequent papers from the early 1980s assure us almost unanimously that cellulosic conversion is "just around the corner" or "a few years from fruition." Taking a positive approach to ongoing research is certainly a good course of action, but small-scale alcohol fuel efforts are hardly legitimized by repeated promises of "feel good" technologies before they are economically feasible on a micro-enterprise level. As fuel prices go up, legitimate would-be small alcohol producers should be prepared to separate the wheat from the chaff so to speak, and approach extraordinary claims with their eyes fully open, especially where an investment is involved.

Forage Crops

Although these are technically cellulosic crops, forage such as Sudan grass, switchgrass, miscanthus, alfalfa, and forage sorghum have potential as feedstock because of the uniqueness of their growth cycle. In the early stages

USDA ARS Photo Unit, USDA Agricultural Research Service, Bugwood.org

Fig 5.6: *Switchgrass has been nearly synonymous with cellulosic ethanol production ever since it was highlighted in the 2007 State of the Union address, but it is not yet practicable for small-scale producers.*

of development, the plants produce only small amounts of lignin and the percentage of carbohydrates in the form of resilient cellulose is relatively low when compared to the fully developed plant. By harvesting these forage crops early, and repeating several times through the growing season, the carbohydrates can be stored in a form suitable for ethanol conversion, and the protein and sugar fermentation residues can also be used for livestock feed. Again, forage crops hold promise for practical ethanol production, but are not at the small-scale level yet.

Crop Residue

Wheat straw, stover (dried stalks) and woody material cultivated for ethanol (such as fast-growing hybrid poplar trees) are primarily cellulose materials. Because of the complexity of the bonds that exist in cellulose, woody matter isn't economically feasible for ethanol production on a small scale, except as a fuel source for heating apparatus. The latest inroads into cellulosic conversion described in Chapter 10 give an overview of the ethanol-from-cellulose process.

Waste, Surplus and More

The small-scale operator who has the curiosity and time to experiment may find a gold mine in wild growth and agricultural or commercial waste products. Many plants are under study for feasibility of ethanol production and the use of surplus, discarded, or even spoiled crops and foodstuffs is nothing new. But the rising cost of petroleum fuels has renewed the interest in these "oddball" feedstocks and some may be especially suited to small-scale production under the right circumstances.

Cattails

These familiar perennial plants grow in wetlands, ditches and marshy areas. A creeping rhizome, or root structure, supports foliage that can reach three to eight feet in height. The rhizome is where the starch is, and it can be harvested for flour and fuel. Cattails have received much attention as means to purify lagoon waste in constructed wetlands. The nitrogen- and phosphorous-absorbing plants thrive in nutrient-rich effluent to remove chemicals, solids and dissolved nutrients that eat up the water's oxygen and defile our natural waterways.

Harvesting the plant for energy is not exactly a new idea. An early proposal in 1979 suggested a yield of 1,500 gallons of ethanol per acre, which would appear optimistic. More recently, in 2008, biofuels researcher Professor Abolghasem Shahbazi of North Carolina Agricultural and Technical State University and his students completed a trial in which they used cattails cut to water level, dried and ground up to make a cellulosic feedstock that, with the use of multiple enzymes and yeasts, was fermented

Biofuels Research Lab, School of Agriculture and Environmental Sciences, NC A&T State University

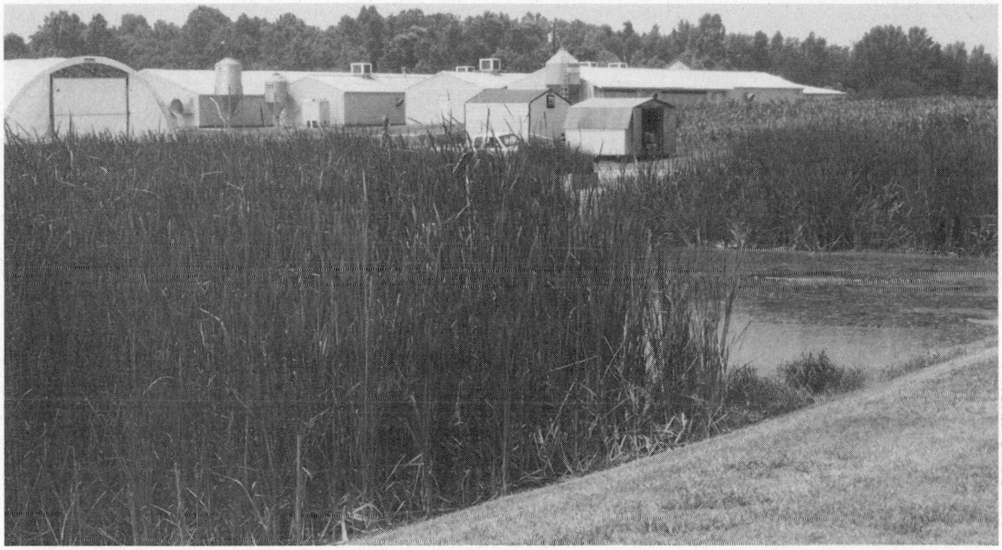

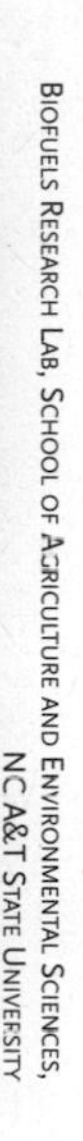

Fig. 5.7: *Cattail research at North Carolina A&T State University.*

and distilled into ethanol. In that trial, 7.2 dry tons of feedstock were harvested from one acre, and the cellulose conversion rate was around 43 percent. Future trials will be optimized for much higher conversions.

Though the actual starch content of the rhizomes can only be speculated upon, figures of 40 and 45 percent have been offered. The leaves also contain some sugar, but yields would likely be low on a weight basis. However, without the ability to cook and hydrolyze the plant starch with conventional enzymes, cattails remain in the realm of cellulosic research.

Cheese Whey

Whey is the thin, watery portion of milk that is separated out as a by-product in the process of cheesemaking. Although the whey contains lactose, or milk sugar, the sugar content is not particularly high — perhaps 5 percent, along with 2 percent protein. The remainder of its composition is essentially water. This presents a problem in fermentation, since the alcohol content in the fermented mash is subsequently reduced to about 2.5 percent, five times lower than a typical ferment.

There are two approaches to get around this. The first is to concentrate the sugar by evaporating the water, but this requires additional energy — not worth the expense unless there was a renewable source of heat available. Even on an industrial scale, evaporation is eschewed in favor of a reverse osmosis process, which separates the water using far less energy. The second method is more straightforward, in which the liquid whey is used to supplement water (see below) in the fermenting process. With corn as a feedstock, 15 percent of its weight can be replaced by whey.[8]

Whey also poses another difficulty in that it's not easy to transport. Not only does it need to be maintained at 130°F in transit and used within 48 hours to prevent spoilage, but its low sugar content means moving a lot of liquid to achieve a fairly low yield. You can expect about 25 gallons of alcohol from 1,000 gallons of whey in a small-scale operation. Realistically, whey may be a good candidate for ethanol production only if the processing facility were part of a cheesemaking operation.

The lactose in cheese whey does not respond well to the yeasts used with glucose or fructose. Only a lactose-fermenting yeast such as *Saccharomyces fragilis*, *Kluyveromyces fragilis*, or *Torula cremoris* will metabolize and convert the lactose sugars. These strains are less alcohol-tolerant than conventional yeast, and are not commonly available, so they must be cultured. As an alternative, a lactase enzyme can be used to break down the lactose to simpler sugars so conventional distiller's yeast can be used for fermentation. That way, the sugar-laced solution can replace some of the water used to dilute corn, fruit pulp and other conventional feedstocks.

Citrus and Tropical Fruit

Oranges, lemons, pineapples, papayas, mangoes and plantains are a few of the tropical and subtropical climate crops affected by spoilage and surpluses. Most are moderate in sugar content, but some, like the plantain, contain starch. These crops may be useful to an entrepreneurial soul who can access them in the field or at packing and processing sites. Generally, it would be more desirable for a small-scale producer to have one reliable source of a single feedstock rather than to experiment with blending or multiple feedstocks — but depending upon the cost of processing that may not always be the case.

A farm-based citrus ethanol operation would seem to be one way of capitalizing on available product. In Florida alone, citrus processing yields about 5 million tons of wet waste annually.[9] If the fruit can be peeled, processing the pulp for fermentable sugars is a natural progression. (Dried peel residue is currently sold as low-value cattle feed.) Citrus can also be fermented whole-fruit, but the peel contains oils which will contaminate yeast. These would have to be cooked out — but the limonene in the oil has some value as a by-product and could be condensed from the evaporate with some planning.

Beverage, Bakery and Candy Waste

Food waste of most any kind is the perfect feedstock for someone who's not in an agricultural setting or has no easy access to farm products. Industrial and food-processing wastes are particularly good candidates because they are generally specialized feedstocks. Just under 30 percent of the food grown in the US is wasted or discarded, so there should be no lack of supply.

Here are some examples to consider: The enriched syrup from soda bottling plants is present wherever there's a bottling facility, and it is quite sugar-dense. Bakery sweepings and pastry droppings are also rich in sugar. Candy waste — either trimmings or damaged product — is very high in sugar content, as you might imagine. Potato chip bakeries produce a lot of damaged chips heavy with starch, and canneries (beets, pumpkin, squash, peas, fruits) can offer all sorts of sugar- or starch-based feedstock possibilities.

While on the subject of fruit, it's an especially good candidate because of its short lifespan. Size and cosmetic appearance also comes into play because both can render a perfectly good crop imperfect in the marketplace and thus ideal for disposal. Short-term storage is a must with surplus fruit and has to be considered, but alcohol production can be coordinated with the supply. Fruit also requires less water than other feedstocks because it has its own juices, and it ferments easily without enzymes. The pulp residue can be composted, or used as fertilizer or livestock feed supplement.

In some cases, it may take some negotiation with the proprietors or managers to work out an equitable arrangement for access to the spoilage, since they may have an animal-feed market already available. But keep in mind that you don't want to enter into a lasting commitment immediately — you'll need to test the feedstock carefully for conversion and production (and consider its transportation, storage and by-product usage) before closing a deal. Under those circumstances, it should be easier to get hold of some sample batches for testing, since the possibility of a tangible long-term arrangement is there.

The Dynamics of Yield

As you may have realized after reading this chapter, calculating firm alcohol yields for any given feedstock is somewhat difficult. This may not be so important if you're just experimenting with a renewable fuel or making a little ethanol for your vehicles, but in a productive ethanol operation or cooperative effort, you'll want to figure out beforehand what your potential production may be.

Clarence Goosen

Fig. 5.8: *Apple pulp residue is a familiar component of livestock feed and can be used as an ethanol fuel feedstock as well.*

I can tell you that the two biggest variables for the novice are crop yield and ethanol yield from a particular feedstock. Academically, establishing crop yield is no more complicated than accessing extension service or USDA data; realistically, those figures are so dependent on local and environmental variables that the best you can hope for is an average which is essentially what you're getting. Since many potential fuel producers will buy, scavenge or barter their feedstocks, they won't ever be involved in agricultural production, so it's less of an issue.

How much ethanol you'll get from a volume of stock is also a matter of an almost overwhelming set of variables. Processing methods, cleanliness of operation, enzyme and yeast choices, temperature management, environmental control, and quality of the equipment will all come into play, along with your experience. The best advice I can offer here is to keep accurate records during production runs, do some research, and learn from your mistakes.

In examining agricultural data related to fuel alcohol, there's an important correlation between yield per acre and yield per ton that can be deceptive if you don't appreciate the nuance. Refer to Table 5.1 and 5.2 and look at rye, for example. It is in the top third of alcohol yield on a per-ton basis, but right near the bottom in yield per acre. A crop's carbohydrate content, or energy density, accounts for this, but it's a significant factor. If you're growing a feedstock for ethanol production on a limited amount of acreage, you'll want a manageable crop with a relatively high yield per acre. On the other hand, if you're sourcing feedstock from elsewhere or looking at culls or spoils, then the yield per ton is far more important (taking cost into account, naturally), since you're not really concerned with the "growing" end of things.

The cost of energy is also a huge factor in determining whether the ethanol you make is worth your effort. The dynamic of Energy Returned On Energy Invested (EROEI) goes beyond the efficiency of the cooker and the still. If you do not consider the cost of transporting feedstock, preparing it for storage if needed, processing it effectively, and storing it without loss, you're either in denial or setting yourself up for disappointment.

And always keep in mind that sometimes the yield of the ethanol is only one piece of the jigsaw puzzle. The value of by-products such as animal feed supplements or a co-product of the preparation or distillation process may significantly offset your costs and make yield less important. The hulls or peels from a feedstock may be a commodity for an industry you're not yet aware of; carbon dioxide is easily salvageable and has commercial value; cultured enzymes and yeasts have potential as a value-added product; process hot water can be recycled to another part of your operation; crop stalks and hulls can be burned or processed into palletized boiler fuel. The list goes on.

Once in a while the mountain must come to Mohammed. In a cooperative venture especially, it can make more sense to locate the distillation equipment right at the food processing facility than to haul the raw feedstock to a separate alcohol plant. In Brazil, where there are hundreds of independent sugar cane mills, small ethanol plants are set up on site to take advantage of juice extraction, which uses the same equipment and process, regardless of whether sugar or alcohol is being manufactured. Other crop-processing plants in the US are using the same model.

SIX

Starch Conversion, Sugars and Fermentation

How fermentation works • Primary and secondary conversion • The role of enzymes • Barley malt • The significance of yeast • Mashing recipes • Milling and preparation • Sugar extraction and fermentation • Cooking • Available alternatives • Testing procedures • Sanitation and bacteria • Temperature control • pH adjustment

This chapter is one of the most important in the book; it covers the "culinary" aspects of alcohol production: preparation, cooking, fermentation, and all the small details that go into making a recipe a worthwhile success. It is not an easy chapter to assimilate, and it is long besides, but after a bit of scrutiny it will make sense. To maintain a consistent and sequential flow of information, I've included working techniques for both starch and sugar feedstocks, even though they are two distinct types of crop. In the end, the fermentation process is essentially the same, but I've pointed out instances where subtleties occur.

Basic Starch Technique

Starch crops differ from sugar crops in the way they use and store glucose, a monosaccharide, or simple sugar. While a mature sugar crop like sorghum cane stores glucose in its stem (along with some starch in the grain head), a starch crop converts whatever glucose is left — after building cellulose and lignin for structure — to starch for storage in the seed (in grains) or the tuber (in root crops).

Starch is just polymerized units of glucose — molecules linked together in long chains. The chains can be hydrolyzed, or broken down into single glucose units, in a series of steps, first to dextrins, an intermediate form of short-chain glucose, then to disaccharides (maltose, sucrose, lactose) and monosaccharides (fructose, galactose, glucose), the simple two- and one-molecule sugars that yeast are able to digest.

The starch can be hydrolyzed in a number of ways — with hot water, hot water and pressure, acids, and with the use of fungal or bacterial enzymes. On a home or farm scale,

Fig. 6.1: *A starch molecule*

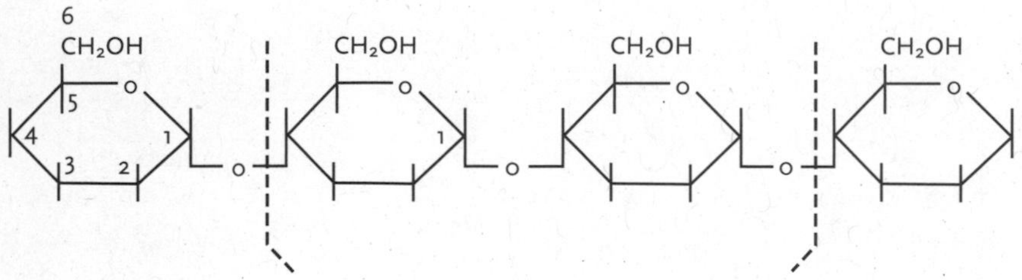

CLARENCE GOOSEN

CLARENCE GOOSEN

Fig. 6.2: *A local mill or co-op can grind grains to the needed consistency for ethanol production.*

some of these options are too costly and complex to be practical, so small-scale starch hydrolysis is usually limited to hot water cooking and malt-derived or commercial enzyme treatments.

The technique can be broken down into five basic steps: (1) milling, (2) slurrying, (3) liquefaction, (4) post-liquefaction, and (5) conversion. The sixth step, fermentation, occurs once the simple sugars are available, and this step is essentially the same whether starch or sugar feedstocks are used. Various ways of describing this process may include a separate hydrolysis stage and a single liquefaction stage, but the technique remains essentially the same. Corn and other crops that contain hard, flinty starches require a two-step process in the slurrying stage that involves heating and boiling; a pre-malting technique that introduces a small amount of alpha amylase enzyme facilitates the slurrying to prevent lumping and over-thickening at higher temperatures. Softer starches like potatoes and rye can be processed at somewhat lower temperatures, and steps can be combined with the use of specific enzymes.

Step 1: Milling

The reduction to starch molecules begins by physically grinding the seeds or kernels into pieces small enough to expose the granules to water and make them accessible to the enzymes. Soaking, cracking or rolling the grain is not sufficient to expose the starch — it must be grist- or hammermill-ground to a medium or fine meal. Finer grinds below 1/8-inch are used in commercial operations where grain recovery is accomplished with sophisticated equipment, but in the event that you want to remove the distiller's grains after fermentation, it is difficult when the material has been ground to a fine particulate. If you are not planning to recover distiller's grains, a finer grind will increase alcohol yields, to a point. Most rural areas have a community mill or feed store where you can have your grains ground for a small charge. If you're serious about ethanol production, though, you'll probably need to invest in a small-scale hammermill or grinder, further described in Chapter 8.

Residual material like pieces of stalk or cob should be screened out of the feedstock before milling. They have no effect on the mash other than to add bulky matter that cannot be converted to glucose with the techniques being described here.

CLARENCE GOOSEN

Fig. 6.3: *Introducing ground corn to a test batch prior to fermentation.*

Step 2: Slurrying

It takes hot water to swell the starch granules enough to get them to rupture and release their amylose (the inner soluble portion of a starch granule) — a process known as gelatinization. The amount of water needed per bushel or weight of grain varies depending upon the type of grain and the process you're using, but the trick is not to add *too* much, to avoid diluting the sugar concentration, which ideally will be in the area of 20 percent. The water may not be introduced all at once (it depends on the recipe) but in any case it should be in a pH range of 5.5 to 7.0, free of chemicals, and potable enough to feed to livestock as "sweet water" after the recovery of distiller's grains.

With corn, a bushel weighs 56 pounds and 65 to 72 percent of it is starch. It will take about 28 gallons — maybe several more — to slurry and then hydrolyze the ground meal. You don't want to risk rapid gelatinization because the viscosity of the wort (the mixture of grain and water prior to the addition of yeast) will become so great that agitation becomes difficult. The way to avoid this is to add the ground grain before temperatures get into the gelatinization range, which begins at 155°F, but can be considerably higher with a greater amylose content.

Start with 20 gallons of water per bushel, slowly add the grain, then bring the temperature up toward the 155°F point. If you've used a portion of spent mash from a prior run, the temperature will be warm to begin with. It's important, also, to begin slow agitation as soon as the grain is introduced, because this maintains a uniform consistency in the wort, and keeps scorching and hot spots from occurring.

With other grains such as rye, barley or rice, the moisture contents are very similar, and the carbohydrate percentage is also close to that of corn. Their gelatinization temperatures vary, but are generally below that of corn.

The pre-malting treatment for corn and other similar hard starches can be executed at this point. Alpha amylase is added in a modest quantity, approximately ½ ounce per bushel of grain, and the temperature is brought up to 190°F or more, while agitation continues. Do not let temperatures drop below 180°F at this stage or the starch may retrograde into a thick mass, despite the enzymes. Within minutes of being introduced, the wort will thin out and will also impart a strong, almost bitter, aroma.

Table 6.1
Gelatinizing Temperatures of Selected Feedstocks

Feedstock Material	Total Carbohydrates Percent	Percent Moisture	Gelatinizing Temperature Degrees F.
Apples	12-15	80-85	—
Barley	50-70	10-12	145
Beans, pinto (dry)	60-70	8-9	—
Corn	65-72	10-15	155
Crabapples	17-18	80-82	—
Grapefruit pulp	10-11	88-90	—
Lentils	60-63	11-12	—
Millet	65-72	10-12	—
Oats	60-68	10-12	—
Potatoes	15-18	75-80	145
Pumpkins	6-7	91-92	—
Rice	70-79	10-12	142
Rye	65-73	10-12	131
Wheat	65-75	10-12	145
Whey	4-6	93-95	—

Continue to apply heat until the mixture comes to a boil; corn will not completely hydrolyze unless it's kept at an active boil for at least 45 minutes. The high temperatures of this cooking process destroy the initial dose of the alpha amylase, but it has already served its purpose. A follow-up dose will be used in the liquefaction stages to come.

Step 3: Liquefaction

Liquefaction (the conversion of starch to a water-soluble semi-complex sugar) is necessary to keep the slurry from gelatinizing into an unwieldy, viscous mass, just as the pre-malting process does for the hard starches. The combination of heat and alpha amylase enzymes, whether they are commercial or malt-derived, will begin to break down the starch into dextrins. At this stage, correct temperature, agitation and pH levels are important to keep liquefaction times to a minimum — not just to hold processing costs down but also to minimize the risk of bacterial infection. A 45-minute period with agitation should be sufficient, and temperatures in the range of 180°F to 190°F will encourage an effective reaction. The pH levels will vary throughout the process, but should remain within the spectrum of 5.0 to 6.5 (commercial enzyme manufacturers will be specific about the temperature and pH requirements of their products).

Alpha amylase enzymes are used at a ratio of about ¾ ounce per bushel, or at recommended dosages based on the dry weight of carbohydrates in the feedstock (see the section on "Testing Procedures" in this chapter). If you're working with a variety of feedstock materials or with spoils and surplus, it may be difficult to determine what the exact percentage of carbohydrates in the feedstock actually is. In these cases, it would pay to conduct small-scale tests by mixing a known amount of dried feedstock material with a known amount of water (at 8.33 pounds per gallon) to make two gallons of test wort.

The wort can be brought to a slow boil over a period of an hour or so, then allowed to cool to around 190°F before adjusting pH. With that done, you can split the batch of wort into five or six equal amounts in separate, clean containers. Calculate the amount of yeast needed for the small volume of wort in each container, and dose one as a control and the others in progressively higher and lower doses, recording the measurements for each. Maintain the containers at the 190°F temperature in a water-filled tub, and keep the samples agitated by stirring them several times over the period of an hour or so. At that point you can cool the samples and perform an iodine test on each one to determine which dosage was most effective in converting starches to sugars. A sugar test with a saccharometer will help to indicate percentages for conversions and sugar feedstocks.

Step 4: Post-Liquefaction

Advances in enzyme development and the benefits of industrial technology have made this step less troublesome for some, but for the small-scale operator, post-liquefaction can be of crucial importance to fully dextrinize any remaining starches in the wort. This can be simply done by introducing a second dose of alpha amylase enzymes to hydrolyze that starch. The temperature is maintained or reduced to the optimal range for the enzyme, and the balance of any additional water is added, if needed, and the pH checked before adding the enzyme at a level of ¾ ounce per bushel, which can vary. Post-liquefaction can take from 15 to 30 minutes to complete.

It's worth noting that the softer starches such as potatoes and barley can be easier and less costly to work with because the liquefaction and cooking process can be streamlined. This is because the soft starches don't have to be boiled (as corn does), but merely need to be cooked at 200°F or so to liquefy. Commercial alpha amylase enzymes will survive at these temperatures for more than an hour and will still function in their preferred range of 180°F to 190°F. With continuous agitation and stringent temperature and pH control, reduced heating time will translate to energy saved.

Step 5: Conversion

The conversion step is also called saccharification, and it begins once the bulk of the starch has been converted to dextrins. (You should realize that 100 percent conversion is probably an unrealistic goal on the small scale — it's likely that a very small percentage of the starch will remain because it hasn't been completely gelatinized.) Now the objective is to convert the intermediate dextrin sugars to the glucose, fructose, sucrose, and other simple sugars mentioned at the beginning of this section.

A second enzyme, glucoamylase (or beta amylase from barley malt) will do this job. The commercial glucoamylase enzymes are probably more effective at the task simply because they act upon a wider range of dextrins, but their cost may offset some of that effectiveness for those who want to maintain a sustainable operation with a minimum of outside supplies. It will take at least ¾ ounce per bushel of grain to make the conversion, and significantly more for some commercial enzymes. (The manufacturer will specify the requirements in its data sheet.)

The first step in conversion is to reduce the temperature of the wort. This can be accomplished by pumping cold water through the cooling coils or simply by adding more water to bring the mix to around 140°F. Too much dilution will affect the percentage of sugar in the wort (which will be somewhere between 16 and 22 percent at the end of the process), but you'll check that again before beginning the fermentation step.

Generally, the temperature should not rise above 150°F or fall below 125°F at this stage, but there are enzymes that remain functional below 105°F. These, along with the development of yeasts that can tolerate slightly higher temperatures, are used in the "cold-cooking" process described in the section on starch crop recipes. Earlier forays into low-temperature saccharification with Biocon (US) enzymes some years ago indicated that a fermentation lag of up to eight hours developed after inoculation with yeast, likely due to the immediate lack of sufficient fermentable sugars in the mash.[1]

After you've adjusted temperature, it will be necessary to test and adjust pH if needed. Do this after you've added any water, because its acidity or alkalinity will affect the wort's pH level. The pH should be about 4.5, or in any case lower than the levels seen in prior steps. Some commercial enzymes and beta amylase may require a slightly lower pH level, but the manufacturer's recommendations and your experience will guide that. Remember that reducing pH levels means adding sulfuric or phosphoric acid, so be prepared to take the appropriate safety measures.

At this point, the enzymes can be introduced to the wort. Depending upon the enzymes you use, the saccharification conversion should take between 30 minutes and several hours to complete. All the while, the wort should continue to agitate slowly. You can check for progress by making an iodine test, as explained in the section on testing in this chapter. The goal is a clear reading or a slightly pinkish tint. If the test shows traces of blue or purple, the starch has not fully converted, and you should allow more time for the conversion to work. After several hours, if a retested sample doesn't come up clear, you'll have to live with the fact that a percentage of the mash is unusable for fermentation.

The finished wort should be reduced to a temperature of around 90°F or whatever is appropriate for the yeast you'll be using in the next, and final, fermentation step. Using cooling coils ensures that you won't need to add cool water, which, again, would likely alter pH values. The wort should test for sugar concentration in the range of 16 to 22 percent, as measured with the saccharometer or a refractometer. Once the goal has been reached, you can stop agitation and transfer the wort to your fermentation vessel.

Fermentation

Fermentation is the natural conclusion to the series of steps I've brought you through, but it's really in a category of its own because once a feedstock, or substrate, has been converted to simple sugars, the fermentation process is pretty much the same whether a sugar crop or a starch crop was used to get there. In the section "The Significance of Yeast," I explain how those microorganisms go through distinct phases over a period of 24 to 72 hours in order to decompose those sugars and manufacture carbon dioxide and alcohol. Here I'll just say that it's a chemical process in which a yeast cell and its associated enzymes absorb glucose, fructose or mannose, and the enzymes and coenzymes take the sugar molecule through 12 intermediate stages of change. Early in the 20th century, the research of biochemists Gustav Georg Embden and Otto Meyerhof isolated this process, which is known as the classic Embden-Meyerhof cycle.

Fermentation takes place in a sealed vessel which ideally would have some method (usually a heat-exchanging coil or jacket) to warm and cool the fermenting mixture, a place to introduce oxygen in the early phases of the process, a checked vent to allow CO_2 to escape without letting air back in, and an agitator to stir the mash if needed. The tank also requires fill and drain plumbing, an access port or lid, and some means to allow cleaning. Chapter 8 gets into more detail about this equipment, and fortunately, small-scale operators have a lot of creative leeway in developing what might work best for them.

Once the wort has been prepared and the ideal temperature range of 85°F to 90°F established, it's moved to the fermentation vessel (in some setups cooking and fermentation will take place in the same tank). At that point, you can introduce the yeast.

There is a method for "pitching" yeast, which I've described in the section "Making a Yeast Starter," but you can add yeast directly to the wort as well. The difference is really just in how long it will take for fermentation to kick off and become really energetic. With a larger dose of yeast and a large volume of pitching solution, the reaction might begin in a few hours; otherwise it can take half a day or more to become visibly active.

Still, even in simply adding yeast to the wort, you have to give the wee organisms a head start. Take three to five gallons of fresh, warm water (90°F to 100°F is ideal) in a manageable container and add three or four ounces of malt (or germinated barley grain). Then add the dry yeast at a rate of about ½ pound for each 100 gallons of liquid wort you wish to inoculate. This is the equivalent of about 2½ pounds for a 500-gallon batch, but thicker or pulpy mashes and those with undissolved sugars will do better with a higher rate of inoculation, up to twice as much yeast.

Once the yeast is introduced, you can stir the mixture vigorously and let it work for 20 minutes or so. Adding air to the mix at this point by splashing or stirring it occasionally will help to propagate the yeast. Before going any further, I want to stress that the water must be clean and free from any cleaning bleach or chlorination, as it will have an immediate and detrimental effect on the yeast cells. If there's any doubt, you can boil the water sample and let it sit uncovered to allow the chemicals to vaporize and escape. More than one distiller has come to unnecessary grief over something as avoidable as using highly chlorinated municipal water without filtering out the impurities beforehand.

After the initial introduction period, you can pour the small sample of yeast starter into the fermenting vat to multiply. A key factor in developing a rapid ferment at this early stage is to introduce free oxygen into the mash. Oxygen encourages growth and reproduction. If you don't have access to a welding supplier for a tank, compressed air will suffice. Bubbles can be injected through a micro-bubbler tube that you can make by drilling a capped section of plastic pipe with rows of 1/16-inch holes. It's attached to the oxygen hose and submerged in the bottom of the fermentation vat. Injection should begin from the initial inoculation for a period of 15 or 30 minutes, after which the yeast will take up to 24 hours to absorb the oxygen and reproduce.

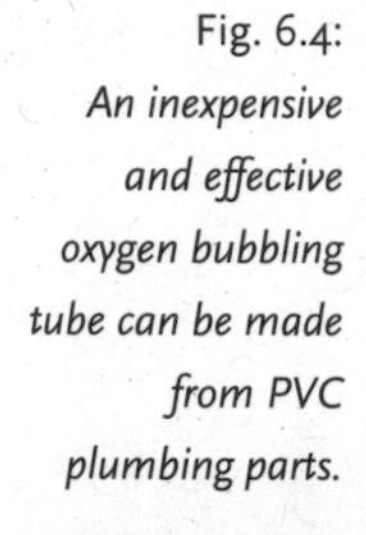

Fig. 6.4: *An inexpensive and effective oxygen bubbling tube can be made from PVC plumbing parts.*

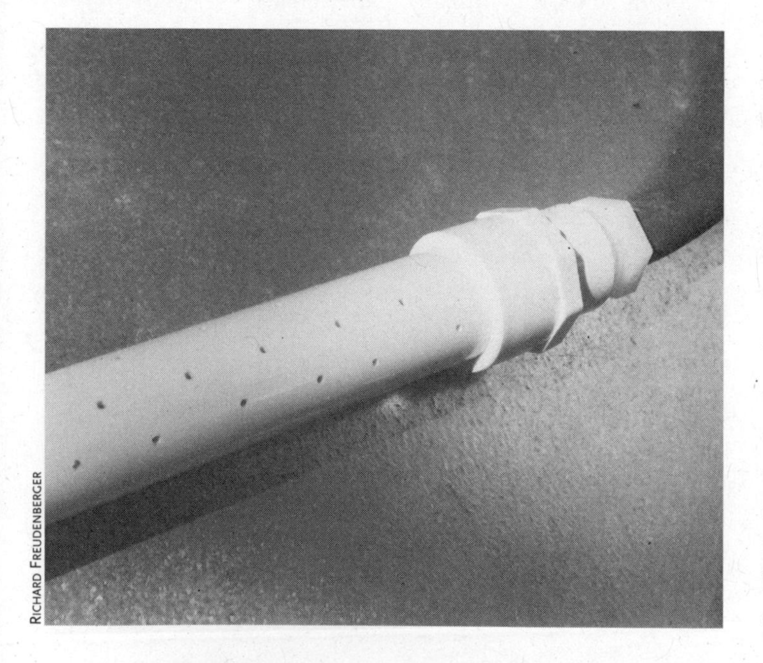

RICHARD FREUDENBERGER

When the yeast moves on to its anaerobic phase, a substantial amount of carbon dioxide will bubble from the fermentation lock, an airlock device that's connected to the vent at the top of the sealed vat. The bubbling indicates that fermentation is working, and heat will be generated as well. With thick mashes, it may be necessary to slowly agitate the mixture, just to assure that all the yeast will come into contact with the nutrient liquid. Too, agitation helps to break up the cap of solids that tends to develop at the surface of the liquid, which can actually choke off yeast from its food supply. Whether you use pump agitation or a blade design, it's critical that you not introduce outside air to the tank in the process, or you'll risk contamination.

After 24 to 36 hours — in some cases up to 72 hours — the bubbling will cease and the fermentation lock will have an odor of alcohol to it. Any floating solids will likely have sunk to the bottom of the vat. At this point, the mash should be distilled immediately to discourage the growth of bacteria, which can acidify the entire batch into vinegar.

The Significance of Yeast

Yeast — the single-cell fungal microorganisms that do the work of converting simple sugars to alcohol — is a far more complicated and industrious organism than you might imagine. There are many dozens of yeast species, but only a handful hold any value to the fuel alcohol producer. To be a good candidate, a yeast strain must ferment rapidly, have a high tolerance for alcohol, be economically practical, and be able to convert the type of sugars present in the feedstock being used.

A variety of dry yeast, or *Saccharomyces cerevisiae,* is commonly used in making industrial alcohol. Distiller's yeast has a higher alcohol tolerance, and specialty yeasts used in highly controlled environments can endure alcohol levels of over 20 percent.

So what happens when yeast is introduced to a food source? Not surprisingly, they mostly eat and reproduce. Of course things are more complicated than that, so we should look at the process in detail. There are two stages of life for yeast. The first is the aerobic stage, in which the environment is packed with oxygen. The yeast take in the oxygen, consume the sugar, multiply in great numbers, and give off carbon dioxide. During this stage very little ethanol is produced, but a lot of CO_2 is generated. The metabolism generates considerable heat as well, enough to warm the environment beyond productive levels — but more on that later.

Once the oxygen supply is used up, the process changes over to the anaerobic stage, where the yeast continue to consume the remaining sugar while producing ethanol and giving off carbon dioxide. As the sugars are metabolized, more heat is generated.

In all, about half the sugar is used to make carbon dioxide, with nearly all the rest being used to make alcohol. A small amount — probably less than 5 percent — is lost to the production of co-products such as acids, adenine phosphates and fusel oils.

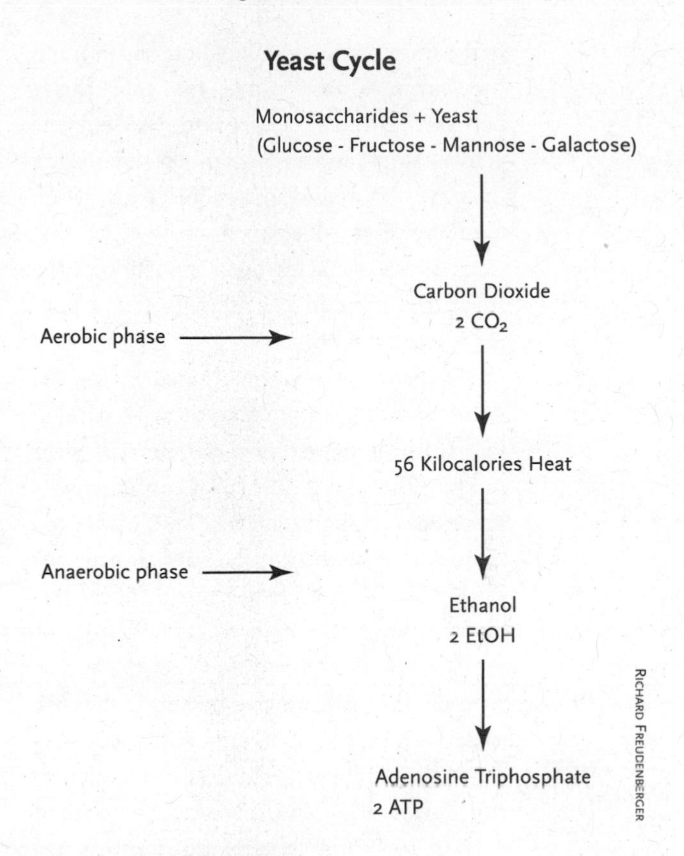

Fig. 6.5: *The yeast fermentation cycle.*

Yeast's Ideal Environment

Sure enough, there are a few conditions that have to be right in order to convert nearly all of the simple sugars to alcohol and ultimately arrive at a mash potency of ten percent or so. They include pH levels, temperature, nutrition and a clean environment. In addition, agitation and ethanol levels must be controlled to keep the yeast working.

pH — Yeast is fairly tolerant of pH levels, but prefers a slightly acidic environment, in the range of 4.0 to 5.0. Moreover, there is still some enzymatic conversion to be done even when the yeast is added (mainly from dextrins to the simple glucose and maltose sugars), and the enzymes responsible for that thrive between 4.5 and 5.5. If pH is too low, enzyme activity stops, and any remaining sugar is left unconverted, reducing alcohol yields. If too high, the yeast will compensate by generating lactic and acetic acids, using a portion of the sugar to do so and leaving that much less to make alcohol with.

Temperature — As mentioned earlier, the yeast's metabolism generates heat — enough to raise the temperature of a 10 percent sugar mash by 64°F. [2] A greater sugar content makes more heat, though some is always lost to the atmosphere with the carbon dioxide or through conduction to the outside of the fermentation vessel.

However, temperature affects the yeast's reproduction rate, which is highest in the 80°F to 90°F range. At cooler temperatures, it takes a long time for yeast to reproduce, though they still work at it. At an elevated temperature of 95°F to 105°F, the fermentation process will stop, and the yeast will begin to die off if higher temperatures persist. The higher the alcohol content of the mash, the less tolerance yeast has for heat, so finding a comfortable balance can be tricky.

Some newer, modified yeast strains have been developed that are more temperature tolerant, but their cost is greater than traditional strains. Generally, it's cheaper just to include cooling coils in the fermentation vessel to establish the ideal temperature. Under the best of conditions, fermentation will occur over a 36- to 48-hour period. If you begin fermentation at a lower temperature to level out the spike of heat that occurs mid-cycle, the timeframe will be closer to 2½ to 3 days. Another control option is to reduce the sugar content slightly by adding up to 25 percent more water to the mash recipe.

Nutrition — For yeast, nutrition is a diet of the right amounts of minerals and proteins. Mash made from grain feedstocks generally do not require any additional nutrients. The spent beer, or "stillage" from the previously cooked batch contains enough nutrition to satisfy the working yeasts' needs in a grain wort. Using 20 to 30 percent spent mash in place of water when preparing a mash recipe will do the trick, in addition to providing a good recycled use for that water and whatever heat is in it.

Feedstocks that are starch-heavy or otherwise without protein or nutrition may require yeast food or malt — germinated barley grains that create and activate amylase and protease enzymes — to hasten fermentation. The section on malting later on in this chapter covers the basic steps of sprouting barley for malt. If additional nutrients are needed, ammonium salts (ammonium sulfate fertilizer) can be added.

Sterile Environment — The last thing you want during — or even directly after — fermentation is to encourage bacterial contamination. This means starting the yeast cultures in a clean environment and keeping the fermentation vessel and equipment in as sterile a state as practical. Achieving a bacteria-free condition is nigh impossible in a small-scale operation, but you do need to minimize contaminant microorganisms for several reasons. The first is that the bacteria will consume sugar that would otherwise go to the yeast. The second is that the bacteria will secrete acids, which not only affect the yeast's metabolism, but also lower the pH of the mash, as noted earlier. These events all reduce alcohol yield, sometimes to the extent of ruining an entire batch.

The simplest way to detect an infection is to monitor pH levels during fermentation. At the start, pH levels of 5.0 or 5.5 are normal, and they will drop a point or so by the end of the process owing to enzymatic action of the yeast itself. If pH levels get down into the 3.0 level or below, it's a fairly good indication that bacteria have gained a foothold.

Cleaning and disinfecting your equipment is a chore, but one that must be done conscientiously. At the start, all containers, plumbing, and of course the fermentation tank itself must be thoroughly washed and rinsed, preferably under pressure. After a few batches have been run, everything must be cleaned again, and scrubbing may be necessary around fittings and heat exchanger components where solids tend to accumulate. Chlorinated cleaners like those used in dairy operations are effective with hot water, as is a lye (sodium hydroxide) solution. Keep in mind that chlorine will destroy yeast organisms as well as bacteria, so it should be neutralized with a hydrogen peroxide solution before the next fermentation run.

Though the chlorinated cleaners are relatively safe, they can still be hazardous if used carelessly. Eye protection and chemical-resistant gloves are considered essential equipment, and respiratory protection may be recommended as well, depending on the materials you're using. The agricultural supply house or dealer you purchase your supplies from should have a Material Safety Data Sheet (MSDS) available from the manufacturer for specific handling and use details.

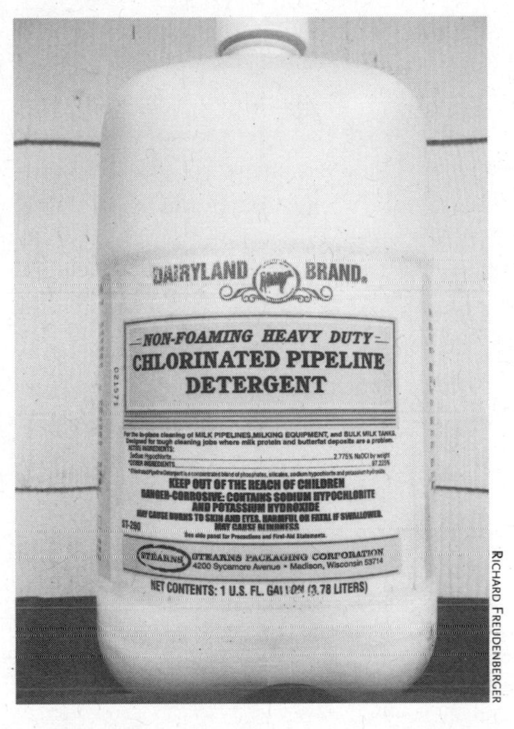

RICHARD FREUDENBERGER

Fig. 6.6: *Standard dairy apparatus cleaning solution can be used to disinfect fermentation and distillation equipment.*

Unfortunately, once fermentation is complete, the risk of bacteria does not go away. The finished mash should be distilled within 12 hours to prevent any active bacteria from making acetic acid — basically vinegar — out of the alcohol and any nutrients remaining in the mash, effectively spoiling it.

Estimating Sugar Concentration

Later, in the section on testing procedures, you'll find that the fermentation process converts only about half of the available sugar from the feedstock into alcohol. And, earlier in this section, we learned that many strains of yeast cannot tolerate an environment *too* rich in alcohol. Given that information, it's apparent that it is in the best interests of an alcohol producer to try and match the sugar levels achieved at the completion of the saccharification stage to the particular strain of yeast being used — or more likely, vice versa.

It would seem, at least intuitively, that providing as much sugar as possible to the yeast population would be a good thing, within limits. But there are several reasons why that approach doesn't wash. Besides the risk of killing off the yeast before they have a chance to consume all the available sugar, starting up initially with high sugar content encourages the yeast to abandon reproductive action and move toward alcohol production. The result is fewer yeast organisms and subsequently longer fermentation times. There's also a greater risk of bacterial infection, since a high sugar concentration slows down the yeast and favors certain types of bacteria.

There is far less risk to having sugar concentrations lower than expected, since the yeast will tolerate the mash and you'll ferment some alcohol, at least. You can also improve both your chances and your yield by choosing a yeast with a high tolerance for alcohol, and by going to the extra effort of making a yeast starter to initiate the wort with a healthy volume of already-working yeast to begin energetic fermentation early on.

Making a Yeast Starter

Though it is possible to inoculate an entire batch of mash with yeast at one time, it's far more effective in the long run to make a separate starter batch of yeast, sometimes called a "pitching mix." ("Pitching" is the term distillers use to describe adding yeast to the wort at the appropriate time; the wort becomes mash once the yeast has been pitched.) If yeast is added straight to the mash vessel, even in a liquid mixture, it could take up to 12 hours for fermentation to begin in earnest, giving bacteria time to establish a footing, and delaying fermentation unnecessarily.

Another reason to make a starter batch is to save on yeast, since the technique reduces the amount needed by about three-quarters. A 1,000-gallon volume of mash inoculated directly requires 12 pounds of pressed (or 4 pounds dry) yeast. It only takes 1 pound of dry yeast to get the job done when the yeast is started and allowed to multiply in a separate batch.[3]

To make a yeast starter, you'll need a vessel or tank that is easy to clean and has a capacity of at least 5 percent of the mash tank. In other words, if you're set up to make mash in 500-gallon batches, the starter vessel should be 25 or 30 gallons in size. As with the rest of the equipment, the starter tank must be clean and free from bacteria that could infect the yeast.

With the mash (or more appropriately, the wort) at a temperature between 80°F and 85°F, pump the 5 percent volume into the starter tank and add any yeast nutrients if required. As mentioned in the nutrition part of the "Yeast's Ideal Environment" section, common grain-based feedstocks probably won't need any additional food; their nutritional content plus the stillage mixed in from a prior mash run should be more than adequate to fulfill the yeast's requirements.

Once that's done, you can add the yeast, keeping the mixture well agitated to assure that there's plenty of oxygen present in the starter environment. (Remember that it is in the initial, aerobic, stage that the yeast multiply profusely, and that's the goal here.) While on the subject, it's worth noting that agitation is an important aid to fermentation throughout the entire process — it keeps the working yeast moving to new sources of sugar and breaks up pockets of carbon dioxide that can get trapped under thick accumulations of material that tend to build up with some feedstocks. The chapter on preparation, fermentation and distillation equipment provides more detail, but a paddle-type blade, an impeller, or even a simple centrifugal pump in the case of smaller tanks will work to keep the mash agitated.

In about six to eight hours, the yeast in the starter tank will have reached an activity level sufficient to support growth in the larger fermentation tank. The contents of the smaller vessel should be added or pumped gradually into the main tank to assure that it's well distributed. Agitation should continue and the temperature should remain in the 80°F to 90°F range.

For large mash batches, it can be impractical to pull off even 5 percent of the wort and use it for a starter all at once. Here, the process is best done in stages, beginning with a small volume — perhaps a gallon or so — of filtered wort. The sample is boiled briefly and yeast is added once the temperature has dropped to 85°F or so.

CLARENCE GOOSEN

Fig. 6.7: *Manual agitation of a small batch of mash during the early aerobic phase of fermentation.*

After an hour of aeration, a second sterile batch is made, this one larger by a factor of 10. When it has cooled to the proper temperature, the first batch can be added to it and the mixture allowed to work under aeration for another hour. This can be continued, increasing the volume of new wort by a factor of 10 each time until the 5 percent figure has been reached. At this point, the starter batch can be added to the main volume of mash.

To make the best use of time, you can begin your starter batch well ahead of actually making the wort. If you're making only one batch of mash and have no prior mash to draw from, you'll have to make your starter from a sterile sugar or molasses solution and a pound of malted barley. You can begin with 25 gallons of water, brought up to the temperature recommended by the yeast manufacturer. Two or three pounds of dry yeast can be initially rehydrated for about 10 minutes by agitating it in a smaller container of 100°F to 105°F clean water (chlorinated water can be boiled first to drive off the chlorine that is toxic to yeast). This makes the cell walls pliable and ready for fermentation — but do not keep the yeast at those elevated temperatures for longer than 20 minutes.

Add the hydrated yeast to the slightly cooler nutrient solution while providing aeration. Then add enough sugar nutrient to bring the sugar percentage to between 15 and 18 percent. Depending upon the yeast, you may have to introduce the sugar in increments to keep the yeast reproducing and not progressing rapidly to the alcohol stage. Allow the starter batch to work at a temperature of 86°F or 88°F for about six hours while aerating it with compressed air or tank oxygen. That time can then be used to prepare your main volume of wort.

The Role of Enzymes

When foods are metabolized, they are synthesized into complex elements, which are transformed into simple elements, which are then converted to energy for use. Enzymes are the catalytic proteins, produced by living cells, that are able to initiate these normal biochemical reactions without themselves being altered or destroyed. Enzymes increase the rate of chemical reactions and are able to do so as proteins, whether they come from microorganisms, plants or animals.

Although enzymes are catalysts, they are also biological components and can be rendered ineffective through exposure to extremes of heat, acidity or alkalinity. Too, there are many different kinds of enzymes, each very specific to its function. Unique enzymes have been developed not only for the ethanol industry, but also for the production of textiles, lumber goods, cleaners, foods and many other products.

Enzymatic reactions are achieved one molecule at a time. In the hydrolysis process, an enzyme combines with the substrate, or feedstock material, at a specific location on the enzyme. Once the match is made, a chemical reaction splits the substrate molecule into smaller components. Then, the enzyme disassociates itself from the substrate, regains any atoms lost in the chemical reaction, and repeats the process with another substrate molecule. Enzymes work very quickly under the right conditions, with activity rates between one thousand and one million cycles per minute. But those conditions have a definite effect on how well the enzymes function, as you'll see next.

The Effects of Environment on Enzymes

Time, temperature, pH levels and enzyme concentration are the key players in how well the enzymes do their job. Let's look at each in detail:

Time—Enzymes function on a steep timeline, breaking down a large percentage of starches to simpler sugars before slowing their activity. Table 6.2 shows that 80 percent of hydrolysis by the enzyme glucoamylase occurs within the first 30 percent of the total activity period. This is significant, for it indicates that the amount of time a substrate is exposed to an enzyme is directly tied to how efficiently it performs. Cutting short the enzymatic activity in an effort to save time in the case above would result in a 20 percent reduction in effectiveness.

A related factor comes into play here as well. Sometimes one enzyme depends upon another to partially break down a substrate before it can do its work; e.g., the two enzymes function simultaneously. When using barley malt as an enzymatic agent, for example, its alpha amylase has to break down the starch into dextrins before the resident beta amylase can reduce them to maltose. Beta amylase cannot reduce starch directly, so the two must have time to work together.

During the liquefaction stage, timing also affects the enzymes' access to feedstock substrate. As heat gelatinizes the starch granules and alpha amylase begins to reduce the starch chain, heat input cannot proceed too quickly or the rapidly rupturing granules form a thick, gooey mass that prevents the enzymes from moving. Once the heat is reduced, gelatinization can slow to the point where the enzymes will regain their mobility and with it their activity.

Temperature—There is a fine line between optimum and excess temperature for every strain of enzyme. While raising temperature to an optimum point increases the rate of reaction, going beyond that point decreases conversion because the heat destroys the enzymes' catalytic properties. Commercial enzymes are specifically developed to work within a defined range.

Temperature and time are closely linked, because enzymes require a certain period to work before the mash reaches the point of thermal inactivity. The mash is usually held

Table 6.2
Hydrolysis by the Enzyme Glucoamylase

Time (Hours)	Conversion Percentage
1.5	20
3.0	40
4.0	50
6.0	70
7.5	80
9.0	85
12.0	92.5
15.0	96.25
18.0	98.12
21.0	98.76
24.0	99.08

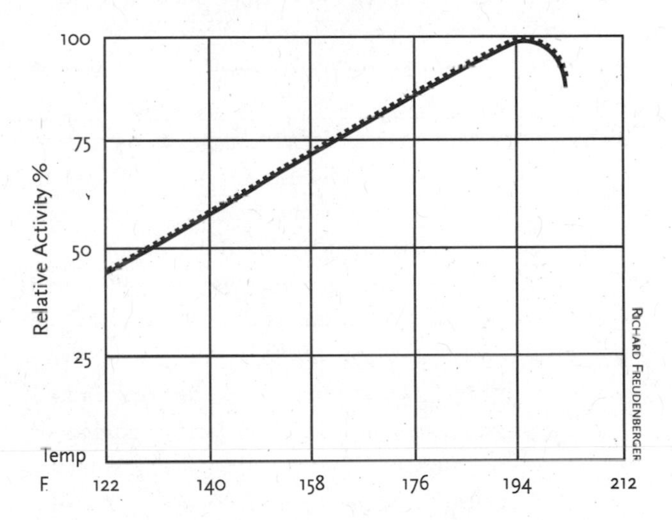

Fig. 6.8: *Alpha amylase enzyme activity increases with a rise in temperature up to about 195°F, where it drops off sharply.*

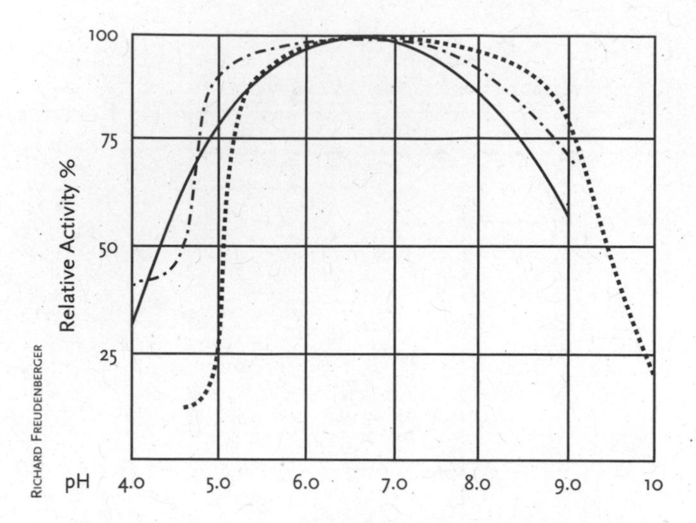

Fig. 6.9: *The optimum pH range for this selection of three different alpha amylase enzymes is between 6.0 and 7.0.*

within the optimum temperature range for a stretch of time before it's allowed to boil.

pH level — Every type of enzyme also has an optimum pH range. Some are broad, with a wide tolerance for pH change; others are very narrow, and subsequently more difficult to work with. When an enzyme is forced to work outside its range, its activity slows and may even stop permanently if held in the unhealthy environment for too long.

Concentration — How much enzyme you use in the mashing process determines to a great extent how much time will be needed for hydrolysis. If, for example, twice the normal enzyme dosage were used, reaction time would be halved. But it's important to note that both the cost of materials and the fact that the final portion of substrate would still take time to convert argue against simply doubling the dosage. Nor will any economy be gained by using half the normal dosage and allowing it to work twice as long. A chemical reaction can only occur if there's enough energy available in the substrate to make that happen. The enzymes only increase the rate of reaction by lowering the amount of energy needed to activate it. It's best to follow the manufacturer's recommended weight-to-volume ratios where dosage is concerned.

Ethanol's Key Enzymes

Grains and starch-based feedstocks for the most part use the amylase enzyme group. These are particularly suited to hydrolyzing starches to different types of saccharides, and the three most important are alpha amylase, beta amylase, and glucoamylase (also called amyloglucosidase).

Alpha amylase is found in animal, plant, fungal and bacterial sources. It's present in all starch products and is formed during the grain-malting process. Its main purpose is to liquefy starch granules during the cooking process and break them down into dextrins, and eventually into maltose and oligosaccharides if allowed to work long enough. Certain bacterial alpha amylase are designed to function in environments of 190°F or greater for cooking purposes.

Beta amylase is another saccharifying enzyme found in microorganisms and plant life. It works in conjunction with alpha amylase by converting dextrins to maltose, but cannot bypass glucosidic branch points (bonds at points of connection) and linkages, so it will leave some of the starch in dextrin form.

Glucoamylase originates from strains of fungi. It's valuable in that it's capable of hydrolyzing both the branch points and linkages of starch, and so can completely convert starch into a fermentable glucose. Its main drawback is that it works slowly, because it hydrolyzes one glucose unit at a time as it moves along the chain.

During the fermentation process, when yeast is added to the saccharified mash, the yeast also produces its own enzymes. Extracellular enzymes function outside the cells that formed them and include maltase and sucrase. Maltase hydrolyzes maltose to glucose, and sucrase splits sucrose to one unit of fructose and one unit of glucose. Intracellular enzymes ensure the metabolism of the cell itself; glucose is absorbed by the yeast cell and fermented, as described in "The Significance of Yeast" section, through what's known as the Embden-Meyerhof cycle.

Making Barley Malt: The Home-Grown Enzyme

Malt is barley grain that has been intentionally germinated for its enzymes. Although other grains can just as easily be sprouted, barley is the grain of choice because it produces a good quantity of the versatile alpha and beta amylases used in the mashing process.

Why, you might ask, would barley be in the business of making enzymes? The answer to that question can be found in the natural cycle of the plant itself. When a dormant seed is stimulated by the right amount of heat, moisture and air, it begins to germinate, and a series of changes come about. The conspicuous ones are the sprouting of the stalk and rootlet; not so obvious is the formation of enzymes that will eventually break down the cell wall between the embryo and the starchy endosperm and, in the process of tearing apart the seed for food and building the stalk and root, convert starch to sugars and back again. When the germination process is halted at the height of enzyme production, the sprouts can be dried and stored or used immediately for their enzymes.

Barley malt isn't a hands-down replacement for commercial enzymes. It has its good and bad points like anything. The latter include the time and labor — and moderate investment — involved in acquiring the seeds, keeping after the sprouting schedule, and processing them for storage. Quality may not be consistent, and the organic enzymes probably aren't quite as effective as manufactured enzymes. But there are significant advantages too: Once you've set up a sprouting operation and gotten some experience under your belt, you will have provided yourself with a truly inexpensive source of enzymes completely removed from the whims and ways of commercial suppliers. You'll have a self-sustaining supply in the control of no one but yourself. If the quality is not always on target, it's a simple matter to use more malt to make up the deficit — and any barley you use adds to the overall yield, since it is a starch source in itself.

Preparing the Barley

The kind of barley used in the brewing industry is a six-row barley, chosen for its flavoring characteristics as well as for the quality of its enzymes. As a fuel producer, you don't care about flavor, but the enzymes remain important. The barley seeds should not be used immediately after harvesting, since they normally lie dormant for a period in their natural state.

Fig. 6.10: *Malted barley contains useful fermentation enzymes for small-scale ethanol production.*

THE BARLEY PROJECT, CROP AND SOIL SCIENCE DEPARTMENT, OREGON STATE UNIVERSITY

Once they've gone through a cold dormant period (three months at temperatures of 60-65°F), they should be sorted to consistent size through a screen. The size itself is not as important as consistency within each batch, because you want to maintain even water absorption. Screening can be done through several sizes to sort the batches as needed. After the seeds are screened and cleaned, they should be soaked in a container of slightly alkaline water kept at approximately 60°F. Assuming you're sprouting a single batch of 50 to 100 pounds of seed, you can use a plastic food storage barrel or small tank. Pour the seeds in and fill the container, then add water and let the seeds soak for 24 hours, while they absorb water. Because air is critical to the malting process, it has to be added to the soaking vat from the bottom through a sparging tube (a length of PVC pipe with rows of 1/16-inch holes drilled through it) plumbed to an air supply, or you'll have to drain and replenish the water every few hours, which is far more labor intensive.

Once the seeds have absorbed enough moisture (45 percent moisture content, or more empirically, when the seeds can be crushed without any hard starch remaining), they can be germinated. One method is to spread them on a concrete floor and turn and water-sprinkle them several times a day to keep them moist, allow the release of CO_2, and discourage bacterial rot through matting. The problem with this is that, even in a warm ambient environment, it may be difficult to maintain a 60°F temperature in a concrete slab without heating it, which could be impractical on a small scale (though the germination generates some heat itself). The process should take place in a darkened or shaded area, so a sun-warmed outdoor pad isn't an option.

The alternative is to dry them on screened racks, similar to the trays used in a large food dryer. They can simply be non-galvanized fine screen stretched and stapled over wooden perimeter frames. The racks should be set up out of the light and in a place where you can mist or sprinkle them regularly in an environment of about 70°F. Again, the seeds must be gently turned several times a day to provide aeration while keeping them moist, and a ventilation fan can help. Another option on a larger scale is to turn them very gradually in a perforated stainless steel drum like the type used in commercial dryers. The drum is set up on a horizontal axis and moist air is forced through the perforations at a controlled room temperature.

The sprouted seeds are ready to use when the small rootlet reaches a length of an inch or so and the immature stalk appears as a bulge along the side of the seed. Once the stalk has erupted more than one-half inch or so, the seed has sprouted too far. To stop further growth and "lock in" the enzymes, you'll have to dry the seeds if you plan to store them. Do this by circulating 120°F air over the sprouts in your trays. Fan-driven warm air works well, but the temperature must be held to no higher than 120°F, because the enzymes can be destroyed at temperatures higher than that. A thermostatically controlled food dryer may be your best option here. Once the sprouts are dry, they can be ground by hand or electric food grinder and stored in a cool dry location.

If you're using the malt right way, you can just grind the seeds green. Fresh malt is actually stronger than dried material, but once the malt is dried, both have about the same potency by weight because the dried sprouts are less dense.

Using Barley Malt

The standard recipe for barley malt is to use 10 to 12 percent, by weight, of malt to feedstock grain. Since the malt contains alpha amylase and beta amylase (the malt's version of glucoamylase), the total volume should be split into two parts to work with the corresponding stages in the fermentation process. The different conditions that exist in each stage prevent both amylase enzymes from working at the same time, though both enzymes are always present in the malt.

In the liquefaction stage, the malt's alpha amylase replaces the commercial alpha amylase quite effectively, although its comfortable pH and temperature ranges are not as broad. The malted enzyme does best at pH levels between 5.5 and 6.0, and at lower temperatures in the post-liquefaction phase, when the second dose of alpha amylase enzymes is added. In the conversion stage, once the starch is converted to dextrins and the pH is lowered, the malted beta amylase enzyme performs similarly to the commercial glucoamylase, even reducing the saccharification period in some cases.

Because the quality of home-malted barley can be inconsistent, it may be necessary to boost the dosage slightly to improve the potency of the malt. An additional 5 percent malt can be used if enzymatic activity appears sluggish, but the total malt-to-mash ratio shouldn't exceed 20 percent.

Testing Procedures

Fermentation, or the anaerobic conversion of sugar to ethanol and carbon dioxide by yeast, is a complex process that takes from 24 to 72 hours to bring to completion. Since every feedstock has a slightly different sugar or starch content — and often both — it's not always easy to determine what percentage of sugar will be available for fermentation after conversion is complete.

The quantity of fermentable sugars in the mash, and more to the point, how efficiently the yeast consumes them, is critical to determining your alcohol yield. As an ethanol producer, it is to your benefit to not only find the best balance of content and economy in a feedstock, but also to make sure that as much useable material in that feedstock is converted to alcohol as possible.

The only accurate way to determine the content of a feedstock is to have it analyzed, which isn't always practical for a small producer (though other avenues exist, as we'll see in a moment, which are far less expensive and nearly as effective). Nevertheless, if you begin with an established volume of feedstock and known quantity of carbohydrates, you'll know what to expect in terms of ethanol yield once the conversion is complete. If the conversion isn't producing what it should, you can then work toward identifying the problem and correcting it.

Analyzing the Feedstock

For the most part, when it comes to choosing a feedstock based on its content, you won't need to hire a laboratory service to analyze the candidates — the heavy lifting has already been done. Years ago, the US Department of Agriculture used its considerable resources — and those of the many agriculturally oriented land grant universities that support it — to research and publish the nutritional value of almost every food crop and product grown or made. The result was the *Handbook of Nutritional Contents of Foods*, a USDA publication that is no longer available in print, but has been published under various imprints such as Peter Smith Publishing Inc. and Dover Publications. More conveniently, the same information is available online at www.nal.usda.gov/fnic/foodcomp/search. Other books listed in the bibliography, including *The Essential Root Vegetable Cookbook* and the classic *Joy of Cooking* contain pertinent nutritional data as well.

In the USDA sources, the foods are categorized broadly into groups — fruits, vegetables, cereal grains, and so forth, catalogued by keyword, i.e., beet, potato, apple, and then further broken down by product, such as raw, peeled, coated, salted, glazed, and so forth. Further parameters are based on weights or serving sizes.

Table 6.3
Water, Protein, and Carbohydrate Content of Selected Food and Farm Products

Crop	Percent Water	Percent Protein	Percent Carbohydrate	Crop	Percent Water	Percent Protein	Percent Carbohydrate
Apples, raw	84.4	0.2	14.5	Okra	88.9	2.4	7.6
Apricots, raw	85.3	1.0	12.8	Onions (dry)	89.1	1.5	8.7
Artichokes, French	85.5	2.9	10.6	Oranges	86.0	1.0	12.2
Artichokes, Jer.	79.8	2.3	10.6	Parsnips	79.1	1.7	17.5
Asparagus, raw	91.7	2.5	5.0	Peaches	89.1	0.6	9.7
Beans, lima (dry)	10.3	20.4	64.0	Peanuts	5.6	26.0	18.6
Beans, white	10.9	22.3	61.3	Pears	83.2	0.7	15.3
Beans, red	10.4	22.5	61.9	Peas, edible pod	83.3	3.4	12.0
Beans, pinto	8.3	22.9	63.7	Peas, split	9.3	1.0	62.7
Beets, red	87.3	1.6	9.9	Peppers, hot chili	74.3	3.7	18.1
Beet greens	90.9	2.2	4.6	Peppers, sweet	93.4	1.2	4.8
Blackberries	84.5	1.2	12.9	Persimmons	78.6	0.7	19.7
Blueberries	83.2	0.7	15.3	Plums, Damson	81.1	0.5	17.8
Boysenberries	86.8	1.2	11.4	Poke shoots	91.6	2.6	3.1
Broccoli	89.1	3.6	5.9	Popcorn	9.8	11.9	72.1
Brussels sprouts	85.2	4.9	8.3	Potatoes, raw	79.8	2.1	17.1
Buckwheat	1.0	11.7	72.9	Pumpkins	91.6	1.0	6.5
Cabbage	92.4	1.3	5.4	Quinces	83.8	0.4	15.3
Carrots	8.2	1.1	9.7	Radishes	94.5	1.0	3.6
Cauliflower	91.0	2.7	5.2	Raspberries	84.2	1.2	13.6
Celery	94.1	0.9	3.9	Rhubarb	94.8	0.6	3.7
Cherries, sour	83.7	1.2	14.3	Rice, brown	12.0	7.5	77.4
Cherries, sweet	80.4	1.3	17.4	Rice, white	12.0	6.7	80.4
Collards	85.3	4.8	7.5	Rutabagas	87.0	1.1	11.0
Corn, field	13.8	8.9	72.2	Rye	11.0	12.1	73.4
Corn, sweet	72.7	3.5	22.1	Salsify	77.6	2.9	18.0
Cowpeas	10.5	22.8	61.7	Soybeans (dry)	10.0	34.1	33.5
Cowpeas (undried)	66.8	9.0	21.8	Spinach	90.7	3.2	4.3
Crabapples	81.1	0.4	17.8	Squash, summer	94.0	1.1	4.2
Cranberries	87.9	0.4	10.8	Squash, winter	85.1	1.4	12.4
Cucumbers	95.1	0.9	3.4	Strawberries	89.9	0.7	8.4
Dandelion greens	85.6	2.7	9.2	Sweet potatoes	70.6	1.7	26.3
Dates	22.5	2.2	72.9	Tomatoes	93.5	1.1	4.7
Dock, sheep sorrel	90.9	2.1	5.6	Turnips	91.5	1.0	6.6
Figs	77.5	1.2	20.3	Turnip greens	90.3	3.0	5.0
Garlic cloves	61.3	6.2	30.8	Watermelons	92.6	0.5	6.4
Grapefruit pulp	88.4	0.5	10.6	Wheat, HRS	13.0	14.0	69.1
Grapes, American	81.6	1.3	15.7	Wheat, HRW	12.5	12.3	71.7
Lamb's quarters	84.3	4.2	7.3	Wheat, SRW	14.0	10.2	72.1
Lemons, whole	87.4	1.2	10.7	Wheat, white	11.5	9.4	75.4
Lentils	11.1	24.7	60.1	Wheat, durum	13.0	12.7	70.1
Milk, cow	87.4	3.5	4.9	Whey	93.1	0.9	5.1
Milk, goat	87.5	3.2	4.6	Yams	73.5	2.1	23.2
Millet	11.8	9.9	72.9				
Muskmelons	91.2	0.7	7.5				
Mustard greens	89.5	3.0	5.6				

Note: From the *Handbook of the Nutritional Contents of Foods*, USDA. Values may differ slightly from online National Nutrient Database values due to continuing research.

So let's suppose you have access to a reliable source of raw yams, as surplus harvest or perhaps as spoils from a nearby processing facility. Let's say you can buy them cheaply, transport them efficiently, and store and process them without difficulty. But will it be worth your while in the long term — in other words, how well will they convert to ethanol?

Referring to the USDA online resource, you would first look under "Vegetables and Vegetable Products" then locate "Yam, Raw" and submit your entry. NDB No. 11601 will appear, also identified by its scientific name *Dioscorea spp.* After you select "100 grams" a chart will come up with nutritional values and the percentage of refuse or non-nutritional material, identified in this case as skin, at 14 percent. In the carbohydrate row, it indicates a value of 27.88 grams — about 28 percent, since the measure is based on 100 units of weight (see the example in Table 6.4). About half of the fermentable sugars will convert to alcohol, so the expected ethanol yield should be in the neighborhood of 14 percent. For 1,000 pounds of yams, this translates to 140 pounds of ethanol. Since we know that ethanol weighs 6.56 pounds per gallon, that calculates to just over 21 gallons of alcohol product. (Keep in mind that this is essentially straight ethanol that has yet to go through the distillation process to separate out the water in the mash.)

In a perfect world, that would be it — end of story. But of course, few things are perfect in life, so you can realistically expect only about 85 percent or so of your potential yield. Quite reasonably, a certain percentage will be sacrificed to the growth and reproduction of new yeast cells, a few more percentage points will be lost to contamination and the inability to pull every bit of ethanol from the mash, and some material will remain as just naturally unfermentable. Given that information, you have an identifiable target. If your effort does not achieve 85 to 90 percent of the potential yield, there is likely some fault in either the mashing or fermentation process, or both.

Only by testing the procedure at different stages will you find out how the conversion is progressing. If you proceed without testing, you're really only making assumptions that everything is occurring as it should; in fact, some processes may not have been fully completed,

Table 6.4
USDA National Nutrition Database Nutritional Example

Yam, raw
Refuse: 14% (Skin)
Scientific Name: Dioscorea spp.
NDB No: 11601 (Nutrient values and weights are for edible portion)

Nutrient	Units	Value per 100 grams	Number of Data Points	Std. Error
Proximates				
Water	g	69.60	12	0.962
Energy	kcal	118	0	0
Energy	kJ	494	0	0
Protein	g	1.53	6	0.268
Total lipid (fat)	g	0.17	6	0.073
Ash	g	0.82	6	0.048
Carbohydrate, by difference	g	27.88	0	0
Fiber, total dietary	g	4.1	0	0
Sugars, total	g	0.50	0	0

Note: The chart has been truncated; Minerals, Vitamins, Lipids, Amino acids, and Other values are not shown.

or the environment may not be appropriate for the enzymes you're using. The result is a low yield and resources wasted.

Testing for Starch-to-Sugar Conversion

Unless you're working with fruits or sugar stocks, chances are that the feedstocks you'll be using are starch-based. The use of the term "carbs" in the nutritional sense may lead to some confusion, since many assume that carbohydrates are synonymous with starch. Actually, carbohydrate is the generic term for organic compounds made up of carbon, hydrogen and oxygen, which include starches, sugars and celluloses as well.

The starch-to-sugar conversion occurs in two steps. The first phase, liquefaction, is accomplished with the use of alpha enzymes (alpha amylase), and converts the long chains of polysaccharides — a combination of 11 or more monosaccharides, or simple sugars held together by glycoside bonds — to dextrins, polymers of glucose intermediate in complexity between starch and maltose sugar. There is little a small-scale producer can do to test at this stage other than make an empirical observation: if the mash is thick and gelatinous after cooking, there may still be a lot of unhydrolyzed starch in the mixture, indicating that this initial phase of conversion is incomplete. Properly done, mash will go through a gelatinization stage in which starch cell walls rupture and the mixture thickens, but will remain fluid throughout the cooking process. This initial starch conversion step must be complete, because the next step takes place under notably different conditions.

The second phase, saccharification, is the final conversion of dextrins to simple sugars, or glucose, by a second series of glucoamylase or beta enzymes. Both temperature and pH are lower in this phase, and must remain so for the enzymes to work. The progress of this conversion can be evaluated using a simple iodine test. Starch reacts with iodine by turning deep purple; the partial presence of starch will appear as blue; a light red or yellow tint signifies no starch, indicating that most or all of the starch has been successfully hydrolyzed.

To prepare an iodine solution for testing, you'll need to buy Lugols Solution, labeled "Strong Iodine Solution USP," at a pharmacy or chemical supplier. Mix 1 ml of this solution with 7 ml of distilled water. As an alternative, you can use common tincture of iodine from the drugstore and dilute it in a 1:9 ratio with distilled water. Both these solutions have a short lifespan and should be used within 24 hours of mixing.

The test is done first by straining a small amount of the wort through a coffee filter to remove any solid particles, producing a near-colorless liquid. Mix 20 ml of this filtered wort with 30 ml of distilled water to make 50 ml of liquid (or any small volume in a 2:3 ratio). Add the iodine solution in tiny increments with an eyedropper, noting any change in the color of the wort sample as you introduce each drop. When the color of the sample no longer changes with the addition of iodine, you've attained a result. As mentioned earlier, a yellow tint represents ideal conversion. Anything deeper, especially toward blue and purple, indicates that there is still unconverted — and thus wasted — starch in the mash.

As tedious as it may seem, it's always a good idea to keep a record of results when you do these tests, to prove consistency. Stick with the same volume of liquids, and log the number of drops used and note the colors observed. This will be especially helpful if you're experimenting with various feedstocks and the enzymes' effects on them to find which conversions are most productive.

In addition, the iodine method can be used to test the effectiveness and dosages of your enzymes, particularly with unfamiliar feedstocks. Enzyme manufacturers optimize their recipes and dosages to suit particular feedstocks, and the small-scale producer may not be working with a common feedstock. Of course, no matter what you were using, you would try to establish the carbohydrate content of your stock as accurately as possible with the resource data available, but it's still only an educated guess. By preparing a small test batch, then setting up a half-dozen or so test sample containers, each with different enzyme dosage levels, you'll be able to establish which works best. One sample should be held as the control, based on the manufacturer's recommended dosage. The others can use proportionately less and more enzyme (be sure to record the dosages). Maintain all the sample containers at the temperature recommended to support enzyme function and swirl the contents of each occasionally to keep it agitated. After an hour or so to give the enzymes time to work, cool the sample containers and proceed with the iodine test, being consistent in the number of solution drops added to each sample. You may discover that it takes more or less enzyme to achieve the best result with your particular feedstock than the manufacturer indicated. Whatever completes the conversion best is the formula you should use.

Testing for pH Levels

The symbol "pH" is used to express effective hydrogen-ion concentration, or in plainer terms, the acidity or alkalinity of a solution. A scale of 0–14 represents the available range, with 7.0 being the neutral value. Any value below 7 indicates acidity; values greater than 7 indicate alkalinity. The pH scale is not linear — it is an inverse logarithmic representation, i.e., each pH unit is a factor of 10 different from the next unit. So, a shift from 3 to 5 represents a change 100 times greater than a shift from 3 to 4.

Because yeasts and enzymes, as living organisms, have fairly well-defined ranges in which they can survive, pH is important. Some ranges can be quite narrow, and if the microorganisms' environment drifts beyond ideal limits, their action will slow down, they will not function effectively, and they may even be destroyed. The enzyme manufacturer should provide you with a data sheet that specifies the storage, handling and use of the enzyme product.

The pH value should be tested at each step of mash production and fermentation. The water used in these processes will, obviously, affect pH levels, but so will the type of feedstock you use, and to some extent your materials-handling equipment. While on the subject of water, it's worth mentioning here that highly treated water, as from municipal supplies, will likely contain chlorine and other additives such as fluoride that may be detrimental to yeast and enzyme organisms. It would be prudent to have a sample of your intended water supply tested by a qualified laboratory to determine if there's a potential problem. The simple fecal coliform tests done by your county health department are not adequate for your purposes. You'll need a more sophisticated analysis covering a range of parameters for elements like sodium, cobalt, and the previously mentioned chlorine.

The national testing labs, or perhaps a state university with an environmental and health program, will be able to provide you with a schedule of the testing levels they have available. A list of courier-service national laboratories is presented in Appendix B. If you're working in a rural or exurban area, it's more likely you'll have access to well or spring water, which may be contaminated with chemical or fertilizer runoff, but not chlorine. Again, a moderately priced water test is a wise investment.

The simplest means for pH testing is with litmus paper. The paper is sold in narrow strips or as a roll, and reacts with a solution by turning color. By matching the hue to a color chart supplied with the paper, you're able determine the pH value at a glance. Indicator papers are sold with the complete 1–14 range, or broken down into narrower ranges such as 3.0–5.5, 3.4–4.8, or 4.5–7.5. The narrow-range papers are more accurate because they make it easier to pinpoint the value.

Using litmus paper is fairly simple. Take or tear off a strip from the container and dip one end into the solution to be sampled. It will turn color right away, and can immediately be compared to the color chart. Be sure to make your match in ample light so you can accurately determine the colors. If you store the paper in direct sunlight (or a damp environment), it will fade (or darken) the test strips and throw off your readings, so keep it in a dry, unlighted environment.

With a bit of expenditure you can circumvent the paper altogether and buy a pH meter. A decent one will cost several hundred dollars or more, but even less-expensive models are more accurate and leave less to chance than indicator paper. The meter has a probe that takes a reading directly from a solution sample and provides a value on an analog scale or a digital readout. Some are calibrated with distilled water, and others with a calibration solution made for the purpose. Any meter worth its salt will have a calibration adjustment. Be aware that the kind of pH meter used to test garden soil isn't appropriate for testing mash solutions.

Once you've established the solution's pH level, it can be adjusted with the addition of lime (calcium oxide or agricultural calcium hydroxide) if it is too acidic, or one of several common acids (sulfuric or phosphoric) if it's overly alkaline. Muriatic, or hydrochloric, acid should not be used, as it can be toxic to yeast. Both acids and bases must be handled using safety equipment, and with deliberate care.

Testing for Sugar Concentration

Since sugar is the basis from which ethanol arises, it is critical to the fermentation process. Whether it's a simple sugar (monosaccharide) or a disaccharide (a combination of two monosaccharide units), it is fermentable. Feedstocks that are mostly starch do not normally require a sugar reading, since what you really want to know is whether the starch has been successfully converted, and the iodine test can be used to determine that. But with feedstock crops like sugar cane, sorghum, and sugar beets —

A Word About Safety

Both acids and alkalines are dangerous substances that render serious and permanent injury if not handled with respect. My own hands are a testament to the hubris of a younger soul who incorrectly assumed that a little bit of lime couldn't possibly hurt the skin of someone inured to axle grease and racing fuel. To this day, a dry winter can be an excruciating reminder of how ignorance — or stupidity — can stay fresh far longer than one could ever imagine.

Basic safety equipment is a must when handling any of these substances. Snug-fitting ventilated chemical goggles are nothing less than a requirement, and gauntlet-style rubber gloves, a long-sleeve shirt, full-length trousers and boots are highly recommended as well. For face protection, an inexpensive facial shield is not out of the question. You can find sales outlets for all of this equipment in Appendix B or on the Internet with a "chemical safety" or "laboratory safety" search. Any university or community college will likely also have an environmental or chemical lab where you can find out more about what you might need.

RICHARD FREUDENBERGER

Fig. 6.11: *Some form of eye protection, gloves and a respirator are part of the safety equipment package. Protective clothing is recommended as well.*

The acids — sulfuric, phosphoric and perhaps lactic if you use it — are best purchased at industrial or chemical supply houses and will be cheaper in volume. It's a good opportunity to make a cooperative purchase with other ethanol producers, and product Material Safety Data Sheets will be readily available from the supplier. A technical-grade acid is sufficient for most purposes, but the purer food grade has fewer contaminants if you intend to use the distiller's grains for feed. Acids should always be added to water or water-based liquids. *Never* pour water into an acid solution; it creates a violent and immediate reaction. Heat also magnifies the reaction of acid, so it should be introduced to warm, rather than hot, solutions. The safest way to add an acid is to pour it into water sparingly while stirring to dissipate heat.

Bases or alkalines like the calcium oxide or hydrated calcium hydroxide you'll come into contact with are less immediate in their effects but still hazardous if used carelessly. The caustic lime (calcium oxide) in particular will draw moisture from the skin and is very irritating to the respiratory system. Lime can be purchased from agricultural or chemical supply houses. As with acid, always use safety gear when working directly with base materials. You can dissolve lime in a small container of water to dissipate it before adding it to a larger volume.

or sugary food waste, or molasses — you'll need to know what percentage is sugar to calculate your ethanol yield.

Sugar concentration can be tested easily with a saccharometer, which is a hydrometer calibrated to a Brix scale. It measures sugar concentration in water based on the density of the liquid. Saccharometers are available in several ranges, from 9 to 21 percent, 18 to 35 percent, or 0 to 25 percent. The bulb of the hydrometer floats in the solution with the scale built into the stem. The reading is taken from the point at which the fluid level meets the stem, which will be partially submerged.

Any solids in the solution will throw off the reading, so it's best to filter the liquid through a paper coffee filter, more than once if needed, to remove any undissolved solid materials. Saccharometers are also calibrated to a specific temperature — 60°F; readings taken above the calibrated temperature will indicate less sugar than there actually is.

Saccharometers are inexpensive, but also a bit inaccurate for the reasons just mentioned. For truly accurate measurements within fractions of a percentage point, you'll need a refractometer, a device that measures the bending of light through a sample of clear sugar solution, using a prism and lenses mounted in a tube (cost: about $200.) The light waves are refracted and projected to the eyepiece, where an etched Brix scale displays a distinct line at the percentage point. Refractometers are also temperature-sensitive, but upscale ATC (automatic temperature control) models even compensate for the difference. If you're serious about making alcohol, the investment in this sophisticated tool will give you better yields by minimizing sugar waste. To put things in perspective, one thousand pounds of feedstock with 18 percent or greater sugar can be expected to yield 14 gallons or more of alcohol.

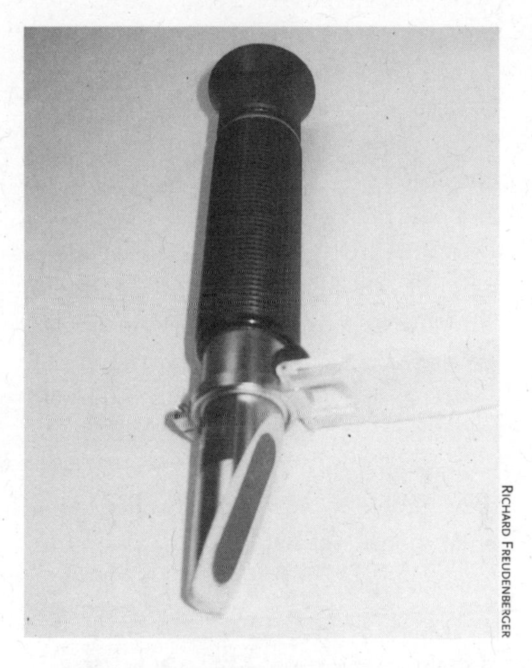
RICHARD FREUDENBERGER

Fig. 6.12: *A refractometer is used to give an accurate measure of sugar concentration in a potential feedstock.*

Testing for Alcohol Content

Your final ethanol product will have to be tested for proof strength, not only to satisfy your curiosity as to how well a job you did, but also to determine the quality of the ethanol as a motor fuel. More importantly, the federal Alcohol and Tobacco Tax and Trade Bureau (TTB) requires that you log the volume and proof of all fuel ethanol you produce.

Proof strength just below azeotropic levels — in the 190 to 192 range — is near perfection for straight-ethanol fuel use. To achieve proofs higher than that requires costly equipment and is, in a practical sense, beyond the capacity of a small producer. Straight-run proofs between 160 and 185 are acceptable as engine fuel — at least they will ignite in the combustion chamber — but water becomes an issue. At a ratio of 80 percent ethanol to 20 percent water (160 proof) fuel mileage is reduced significantly. Of more concern is the corrosion factor due to electrolysis, especially in the presence of zinc and aluminum, two metals common in engine components. Cool-start problems and wintertime crystal freezing also become problematic in lower-proof fuel. All these problems effectively dissipate when proof strength reaches 185 or so.

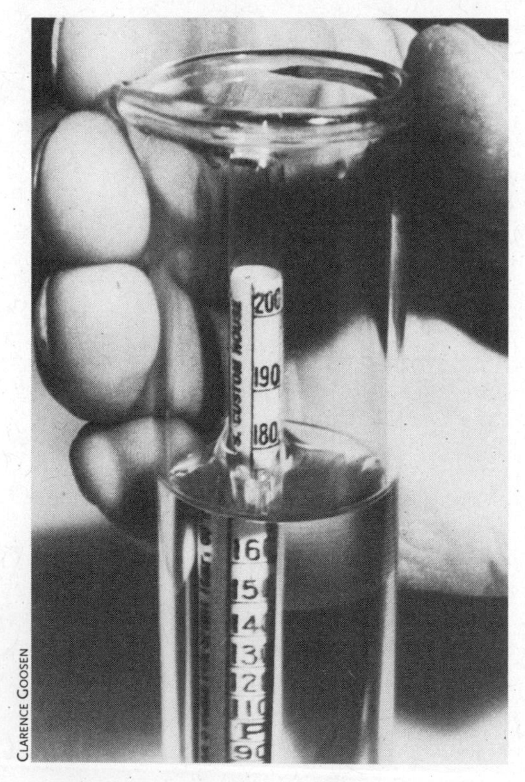

Fig. 6.13: *A proof hydrometer measures alcohol content of a solution, when corrected to 60°F.*

CLARENCE GOOSEN

Like the saccharometer, a Proof & Tralles scale hydrometer measures the specific gravity of a solution, in this case alcohol and water. Proof is calibrated on a 0 to 200 scale, and the Tralles scale represents a percentage from 0 to 100. The proof hydrometer is also calibrated at 60°F and will give a false reading at temperatures above or below that. Higher fluid temperatures result in higher proof readings, so a "True Percent" correction chart published by the TTB (see Appendix A) is used to calculate actual proof at any given temperature between 61°F and 100 °F. This does away with the need to cool every test sample taken from the still.

It is also possible to measure proof strength through a refractometer, but the Brix scale number then must be converted to a Refractive Index number, and then to actual proof. To simplify things, there are refractometers available that are calibrated to use a Refractive Index scale in place of the Brix scale.

A Critical Look at Starch Crops

Even though starch crops are a more complex form of sugar than sugar crops such as sugar beets or sweet sorghum, they are still popular for making alcohol because they're abundant and they store well. Corn (and other grains), potatoes and sweet potatoes are common and can be economical if sourced wisely, and the process used to break down the long molecule chains that bind the glucose together in starch is relatively simple.

Corn, particularly, has been the feedstock of choice for industrial alcohol manufacturers, not because it's particularly the best crop for

CLARENCE GOOSEN

Fig. 6.14: *Corn kernels, a typical production ethanol feedstock.*

energy, but because it's expedient: so much money and so many states are tied into corn production that it would appear almost sacrilegious to use something else. Moreover, to be objective, corn does have a high starch concentration — giving it an excellent yield per ton — and it stores well.

The reasons it may not be the best crop to hang one's energy future on have more to do with scale than production. Corn demands a great deal of water, pesticides and petro-fertilizers, and it requires major capital investment in sophisticated harvesting equipment. It can also be very depleting of the soil, both in minerals and by erosion, if not managed properly. But we do have an abundance of corn, and until recently it has been relatively cheap.

The move toward renewable fuels in general and ethanol in particular has increased demand to the point that the price advantage is rapidly fading, and with it the overall benefits of corn as a fuel crop. Unfortunately, things on an industrial scale don't tend to absorb change quickly, so corn will continue to be a commodity for fuel, as it is for cattle feed and fructose sweeteners.

Starch Crop Fermenting Recipes

On a small scale, corn may not present such issues locally, especially if it's available as surplus, or grown and harvested less intensively. What follows is a conventional recipe for fermenting corn and other grains which we used successfully in our alcohol fuel program, using at the time the Biocon (US) enzymes Canalpha[R] (alpha amylase) and Gasolase[R] (glucoamylase), and the company's Special Distiller's Yeast (*S. cerevisiae*). (Biocon (US) Inc. has since become part of the firm that manufactures Novozymes; the Biocon (US) enzymes were packaged in 2.2-

ounce measures and the yeast in 8.8-ounce packets.)

Conventional Corn Mash Recipe

MILLING AND PROCESSING

Chaff, or pieces of cob and stalk, should be removed from the grain before grinding. The milling can be done with a hammermill screened to produce a coarse to medium grind (not a fine flour grind).

SLURRYING

Add 20 gallons of clean water per 1 bushel (56 pounds) of ground grain. Bring the temperature to 150°F. Begin agitation and add the grain slowly, to minimize clumping. Adjust the pH to the range of 5.5 and 7.0.

LIQUEFACTION

Add 1 packet of Canalpha for each 5 bushels of ground grain. The powdered Canalpha

Hot Cooking: How Conventional Hydrolysis Works

The conventional dry-grind ethanol method uses a hammermill to mechanically break down the corn kernels to expose the starch. In larger operations, destoners and magnetic separators remove pebbles and metallic materials prior to grinding. The milled grain is then piped to a cooker, where it's slurried with heated water to destroy any harmful microbes and airborne bacteria. This is the liquefaction stage, where an initial series of alpha amylase enzymes are added to the slurry after the pH of the liquid is adjusted to between 4.5 and 7.0 to suit the particular enzyme being used. The amylase enzymes must be added before temperatures reach 155°F, the point at which the starches begin to gelatinize, forming clumped masses that are difficult to separate.

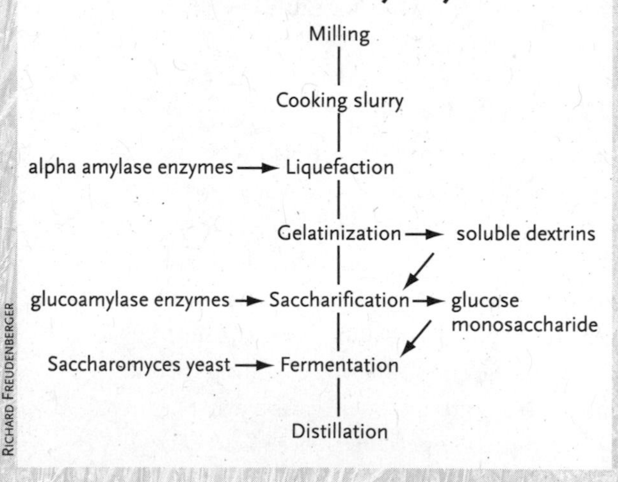

Fig. 6.15: The hydrolysis flow chart.

Typically, about 18 gallons of water are required for each bushel of grain used at this stage. So, a 50-bushel recipe will call for 2,800 pounds of corn and approximately 900 gallons of water. Chlorinated water from a municipal source is treated to remove chlorine that would affect the enzymes. The mash is agitated while heat continues to be introduced to the mixture; usually the mash is boiled for about 30 minutes to prepare the mix for enzymatic breakdown and ultimately hydrolyze the mash, or turn it to an aqueous solution of soluble dextrins. In commercial operations, pressure and additional heat is often used to shorten this time to a few minutes ☛

must be premixed to a paste with 100°F water, then further diluted to a thin liquid before introducing it to the wort. Over a period of 30 minutes or more, raise the temperature to boiling and allow the wort to boil for 15 to 30 minutes.

Post-Liquefaction

Cool the mash to 160°F by adding 5 to 10 gallons of cool water per 1 bushel of ground grain. (The manufacturer recommends 25 to 30 gallons of water per 1 bushel of ground grain as the final ratio.) Add 2 packets of Canalpha (again, first made into a paste) per 5 bushels of ground grain and hold for 30 minutes.

Conversion

Cool the wort to 90°F using cooling coils and add 1 packet of Gasolase and 1 packet of Special Distiller's Yeast per 5 bushels of ground grain.

— yet the cost in energy is dear. The gelatinization phase — during which the successful conversion of starch to soluble dextrins occurs — can take several hours at temperatures approaching 200°F.

When the liquefaction stage comes to an end, the temperature is brought down by active cooling, or is simply allowed to drop to the 140°F range, and the pH is checked and adjusted again if necessary — to around 5.0 — with the use of a dilute acid or a phosphate buffer. At this point a second series enzyme, or glucoamylase, is introduced to initiate saccharification. This enzyme converts the dextrins to glucose, thus preparing the mash for fermentation. Maintaining a constant temperature between 125°F and 140°F along with continual agitation is essential to the action of these glucoamylase enzymes, and promotes timely yeast conversion in the fermentation stage.

Once the conversion of dextrine to glucose is complete, the mash is cooled to 80°F or 85°F, often by adding water. A limit of about 28 gallons (including the 18 gallons used in the slurry) per bushel of grain is the acceptable norm.

Temperature, pH balance, and the presence of nutrients all affect how well the yeasts convert sugar to alcohol. This fermentation process, called zymolysis, is not 100 percent efficient because the yeast uses a certain percentage of the available sugar to create new cells. The *Saccharomyces* yeasts used in ethanol production can use most all of the sugar in solutions between 16 percent and 22 percent while generating a beer containing up to 12 percent ethanol by volume.

Because excessive aeration encourages the yeast to reproduce faster (which is desirable in the initial stages of fermentation), a small amount of liquid mash is mixed with yeast at the start, added to the wort, or unfermented mash, and then agitated briefly to populate the mixture with propagating yeast. After that, the wort is left alone to allow the yeast to work under anaerobic conditions, which promotes the production of waste, i.e., ethanol and carbon dioxide, rather than reproduction, in the microorganisms. The fermentation process typically takes three to five days to complete its cycle, at which time CO_2 production stops and solids drop to the bottom of the container. ■

The powdered Gasolase and the Special Distiller's Yeast must be premixed to a paste with 100°F water, then further diluted to a thin liquid before introducing it to the wort. After the wort is well mixed, stop agitation and transfer the mash to the fermentation tank, if used.

Fermentation

The mash will ferment over a period of two to three days, and will then be ready for distillation.

Simultaneous Saccharification and Fermentation "No Cooking" Corn Mash Recipe

Further developments in enzyme technology have resulted in granular starch-hydrolyzing enzymes such as the type developed by Genencor International. The company's Stargen™ 001 enzyme contains alpha amylase and a glucoamylase that hydrolyze granular starches in corn and other starchy feedstock grains without going through the gelatinization or liquefaction stages (though wheat, barley and rye grains may require a pre-fermentation enzyme step to reduce mash viscosity). The benefits include streamlining the fermentation process and having reduced energy input while still maintaining high ethanol yields. The following section presents a typical Stargen™ 001 recipe for using corn.

Milling and Processing

Mill and then pass cleaned kernels through a 30 mesh (.023-inch) screen. The Stargen™ 001 specification sheet gives detailed information on the effects of alternate mesh sizes; finer particle size gives higher alcohol yield, but makes grain recovery more difficult. Small-scale operations may have to work with a coarser, .125-inch (1/8) screen.

Slurrying/Hydration

Add 40 gallons of 68°F to 104°F water for every 100 pounds of ground grain. Agitate thoroughly.

Saccharification and Fermentation

Add enzyme while maintaining agitation. Use a minimum of 2.2 pounds of Stargen™ 001 for every 2,200 pounds of ground grain (1.6 ounces per 100 pounds of corn). Dosages of up to 5.5 pounds per 2,200 pounds of grain (4 ounces per 100 pounds) may be needed in some cases. Adjust mash pH to 4.0 to 4.5.

Add distiller's yeast or a proprietary yeast with a higher degree of alcohol tolerance (>15 percent) within 15 minutes of pH adjustment. Continue agitation and maintain mash temperature of 90°F. Fermentation will be complete within 90 hours. (Temperature requirements can range between 86°F and 104°F, and fermentation times can vary between 24 and 90 hours, depending on conditions.)

Potato Stock Mash Recipe

Raw potatoes have some characteristics that make them an ideal feedstock. They have high (nearly 80 percent) starch content, high water content, and store well. Their starch granules are large and easily hydrolyzed, aiding in sugar conversion. They are also low in other solids, reducing the need to reclaim spent mash.

The main drawback is their cost, which can be offset by using agricultural or processing culls, or by growing them within your own operation. Another difficulty is in processing: the starch particles must be physically broken down (heat alone will not do it), and the cellulose fibers within the crop tend to absorb liquid and should be roller-squeezed, or "dewatered" to get the fullest amount of wort possible. The following recipe was provided by former co-worker Clarence Goosen, drawn from his extensive research into ethanol during and after the alcohol fuel program we worked on together. The enzymes Taka-Therm[R] and Diazyme[R] refer to products manufactured by Miles Laboratories, Inc., which was absorbed by Bayer AG in 1995. They are interchangeable with any suitable alpha amylase and glucoamylase enzyme.

Grinding and Processing

Grind the potatoes into the smallest particles possible. If you're processing them through a pump as part of the pump agitation cooker (see Chapter 8), limit the size to no greater than 1/8 inch to avoid overloading the rotor. As the feedstock begins to gelatinize, the centrifugal pump rotor will break the particles into increasingly smaller pieces until they are effectively disintegrated.

Slurrying

Potatoes are about 80 percent water, so very little additional liquid is needed to promote flow in the cooker. Add 1 gallon of clean water to every 80 pounds of raw potato. This will yield a 13 percent sugar solution in the final mash. Adjust the wort to a pH level between 6.0 and 6.5, using a 10 percent lime-slurry solution.

Liquefaction

Add 0.5 ounce Taka-Therm (alpha amylase) for every 100 pounds of potatoes in the wort. Maintain continuous agitation and bring the temperature up to 145°F as the starch begins to gelatinize. Continue adding heat at a gradual enough rate that the wort remains fluid. If heat is added too quickly, the starch will thicken and become too viscous to agitate. Should this occur, reduce heat, and let the enzymes gelatinize the starch further. Then continue heating until the temperature reaches 194°F and hold it there for 30 to 60 minutes.

Test for sufficient liquefaction by performing an iodine (starch) test on a 1-ounce sample of wort. Introduce a drop of iodine, stir, and dilute with 10 ounces of clean water. The presence of any dark blue or purple particles in the solution means that the starch is not wholly liquefied and starch chunks have not broken down enough. If this is the case, continue cooking until there are no longer any dark particles apparent in the test.

Remove the heat source and add an additional 0.5 ounce of Taka-Therm alpha amylase enzyme per 100 pounds of potato for post-liquefaction.

Conversion

With liquefaction complete, bring the temperature of the wort to 140°F by cooling it with a heat exchanger or internal cooling coils. Do not add cold water at this stage. Adjust pH to 4.2 with sulfuric acid, then add 0.75 ounce of Diazyme (glucoamylase) for every 100 pounds of potato in the wort. Hold the temperature to 140°F for one to two hours.

FERMENTATION

Cool the wort to 86°F and separate the liquids from whatever solids remain. Add yeast and maintain the temperature at 86°F. Fermentation will be complete in 48 to 72 hours.

Basic Sugar Technique

In Chapter 5, I explained how the non-complex sugar structure in certain crops lends itself readily to fermentation without having to go through the hydrolyzation process. These so-called sugar crops, among them sugar beets, cane, fruits and sweet sorghum, can make excellent low-cost feedstock if processed as soon as possible after harvest to avoid spoilage issues.

As with starch, the sugar technique can be broken down into steps: (1) extraction, (2) cooking, (3) cooling, and (4) adjustments of sugar concentration, pH and nutrients. But cooking or sterilizing the wort can be costly in large volumes, so an alternative is to use a low pH level combined with a high dosage of yeast to knock down unwanted bacteria that would normally be controlled through sterilization. Let's look at all this in detail.

Step 1: Sugar Extraction

There are a number of choices open to you if you're considering a sugar crop as a feedstock. Since sugar feedstocks generally don't keep well for long periods, they should be distilled as soon as they're prepared, and they are usually prepared on a seasonal basis, near or at harvest times.

Traditionally, sugar crops are "juiced," by crushing or pressing them in a machine. Sugar cane is typically crushed in a drum crusher or roller mill, which squeezes out and collects the juice and sends the stalks out a conveyor chute for drying. The juice method yields a high sugar concentration because the liquid is already concentrated and doesn't need any additional water. One drawback of this technique is that as much as one quarter of the sugars available in the plant are left behind in the pressings — but the sugary residue also has value as fertilizer or animal feed, particularly with fruit crops.

Crushers are manufactured in all sizes. They can also be found as used agricultural equipment or as salvage. An auger press is just another, more sophisticated version of a crusher, but these types of presses are usually limited to industrial use and can be quite expensive. A simple arbor press may actually be the best option for a small-scale do-it-yourselfer. The hydraulic types are 25-ton cylinders mounted vertically in a sturdy frame with the piston directed downward. It is meant to press out (or in) bearings, pins and sheaves in machinery, but can easily be adapted to squashing pulp or crop by welding a broader head onto the piston and fitting a sturdy receiver beneath it. The receiver, or container, is fitted with an outlet or screened bottom to capture the juices. The simplest hydraulic presses are manually operated; larger ones have electric pumps and hydraulic valve controls, and use two-way cylinders so both extension and retraction are powered.

There is also a juice method called diffusion, in which pulp or thin sections of crop are heated to 190°F and steeped in solution within a container to draw the sugars from the substrate material. The enriched solution is transferred through several stages of heating and steeping in this manner to progressively draw the sugar from a fresh substrate supply in a series of containers. When the initial supply is exhausted of its sugar, it is removed for use as bagasse or livestock feed and fresh feedstock is added. New substrate is introduced on this rotating basis until the supply is exhausted.

An entirely different approach, and one that is also very feasible on a small scale, is to shred the feedstock whole in a shredder or hammermill. Although the pulped material is more problematic in handling, all of it is used, and the spent mash can still be used for livestock feed. Because some water must be added to the pulp to make it liquid enough to pump or stir, the initial sugar concentration is a bit lower than that of straight juice, but still quite acceptable. A pulped wort like this may benefit from a higher yeast dosage and good agitation because the yeast is not as mobile in the thick solution. Mid-sized shredders are common in landscaping and garden applications and are not nearly as costly as an agricultural hammermill.

Step 2: Cooking ... or Not

Cooking is an option that might best be decided by the economies of your heat source. In fact, one of the benefits of using a sugar feedstock is that you can eliminate the hot-cooking step, as I mentioned just before, to save on fuel costs. But let's assume you're using wood or biomass heat, and choose to cook your wort. For successful sterilization, wort

RICHARD FREUDENBERGER

a

RICHARD FREUDENBERGER

b

Fig. 6.16: *Even a small shedder can be useful for feedstock preparation. The shredding teeth, seen in detail, grind and separate even tough feedstocks.*

temperatures must be held above 200°F for as much as an hour while agitation occurs. The high temperatures destroy harmful bacteria and also help to break down any pulpy material if you've shredded your feedstock whole rather than pressing it.

It's possible to avoid high-heat cooking by simply warming the wort to 90°F, and lowering the pH level — if needed — to 4.0 or 4.5 with the addition of lactic or sulfuric acid. Lactic acid will support the growth of yeast and still inhibit the growth of harmful bacteria. Once the pH and other adjustments listed below are made, you can then introduce a large inoculum of dry yeast (more than one pound per hundred gallons), which effectively competes with the bacteria for food. The larger dose assures — in theory at least — that the yeast will prevail in this competition.

It's important that temperatures not exceed the 90°F tolerance level of the yeast. The ideal range for yeast is between 78°F and 86°F, but ambient temperatures affect this situation as well. In an environment below room temperature, a 90°F wort is desirable, since some heat will be lost to the atmosphere. In a warm climate, the wort may have to be cooled to 65°F or so (see below) to make the best environment. The yeast's metabolism will create heat in any case to raise the temperature.

Step 3: Cooling

If you've cooked your wort, you'll have to cool it before adding yeast. Again, the point is to get it into a range comfortable for the yeast, 85°F to 90°F. Do not add cold water to cool the mix, since that would dilute the wort and almost certainly change the pH level. It's better to use a tubular cooling coil or heat exchanger in the vat, which will circulate cold water and drop the temperature without introducing more water.

Step 4a: Sugar Concentration

The sugar concentration of the wort needs to be adjusted to a range between 10 and 20 percent. Though it would seem that a high concentration of sugar would ultimately yield more alcohol, that argument is only valid up to a point. What happens, in fact, is that the alcohol level in the wort surpasses the tolerance of the yeast, actually altering its structure and curtailing its activity, and ultimately preventing all the available sugar from being converted to ethanol.

A low sugar concentration, on the other hand, makes for an uneconomical use of volume in fermentation and extends distillation time, both of which cost money in the long term. One benefit to keeping sugar concentration at the lower end of the scale (in the 12 to 14 percent range) is that it ensures that most all of the sugar will be fermented. It also protects against overheating due to the yeast's energetic metabolic action in a high-sugar environment.

With shredded, pulpy feedstocks, it's possible to go for a higher sugar concentration by limiting the amount of water added in preparing the wort. (Juices typically have higher sugar content than shredded feedstocks because no water is added beforehand.) Since the yeast can only work on the sugar available to it, it's safe to aim for a higher sugar concentration with pulp feedstock because, initially, much of the sugar is still tied up in the pulpy shreds. As the yeast does its work consuming sugar, more sugar is released over time, but not all at once, so the yeast isn't being inundated with saccharides. Sugar concentrations of up to 22 percent are generally safe with pulpy feedstocks.

A fairly accurate sugar concentration can be gauged with a saccharometer, or better yet, a refractometer (see the section "Testing for Sugar Concentration"). Any undissolved solids in the wort can affect sugar readings, so it's best to filter your sample through an unbleached paper filter before testing it.

Step 4b: Adjusting pH Levels

I've covered the topic of pH in Step 2 above, and elsewhere in this chapter under "Testing for pH Levels." The common acids used to lower pH in an overly alkaline solution are sulfuric and phosphoric, with lactic acid also a possibility. If the wort is too acidic, a base like agricultural lime or caustic lime (calcium oxide) must be introduced. The goal is to achieve a pH between 4.0 and 4.5.

To gauge how much supplemental acid or base to use, withdraw a small sample of your wort, perhaps three gallons from a 300-gallon batch. Take a litmus test or use a pH meter to get a reading, then carefully measure how many units of acid or lime is needed (in liquid milliliters or dry grams) to correct the solution to the proper level. You can then extrapolate this amount to match the total volume of wort. If it took 60 grams of lime to correct three gallons of wort, it will take 5,940 grams (5.94 Kg or 13 pounds) to adjust the remaining 297 gallons. (60 x 297 = 17,820, divided by 3 = 5,940.) Testing first on a sample protects you from "guessing" by trial and error on the whole batch.

Step 4c: Adding Nutrients

Many sugar proteins do not contain all the mineral nutrients needed for yeast metabolism. Yeast foods, in the form of ammonium sulfate or phosphate, are usually added to wort to aid in complete fermentation. These ammonium salts contain phosphorus and nitrogen required by the yeast, and two or three pounds for every 500 gallons of wort is all that's needed. Calcium sulfate (gypsum) fed at the same rate as the fertilizer salts provides needed calcium. Other nutrition additives include prepared yeast foods, but a cheaper alternative is to simply add a five-pound can of dry brewer's yeast to the mix; it will provide nutrients beyond mineral content.

Elsewhere in this chapter I've mentioned using up to one-third spent mash or stillage from a previous run to make up the volume of a fresh batch. This will provide a source of yeast for the fresh wort, but there is also some risk of contaminating the entire batch with bacteria if the spent material isn't used directly after distillation. If there's any appreciable time lag, bacteria will have a perfect opportunity to proliferate.

Pitching the Yeast

Once the wort is cooled and all the adjustments have been made, yeast is introduced to begin fermentation. Unlike the massive dosage of yeast described in the "uncooked" version in Step 2, a less aggressive dose is needed for sterilized wort because it doesn't have to compete with bacteria for food. The typical method for "pitching" yeast (described fully in the section "Making a Yeast Starter") is to withdraw about 5 percent of the sterilized wort, inoculate it with yeast at a rate of about ½ pound dry yeast for each 100 gallons of liquid, agitate it, and allow it to develop a healthy, fast-growing colony before re-introducing it back into the main wort.

An important factor in developing a rapid ferment in the early stages is to introduce or inject free oxygen into the wort. The oxygen encourages growth and reproduction, and it can be easily introduced through a micro-bubbler tube made from a section of pre-drilled plastic pipe attached to an oxygen hose and submerged in the fermentation vat. Compressed oxygen from a welding outfit is ideal for this; using air under pressure is a cheaper alternative, but air does not have as high an oxygen content. Injection only needs to occur in the first 15 minutes or so, after which the yeast will take up to 24 hours to absorb the oxygen and reproduce. Following that, the yeast colony shifts to its anaerobic phase, and substantial carbon dioxide (and alcohol) will be produced, and CO_2 will bubble vigorously from the fermentation lock, a device that prevents air from being drawn into the sealed vat.

When fermentation is complete after 24 to 36 hours or more, the bubbling will cease and the mash should be distilled immediately to prevent the growth of bacteria, which can turn the mixture to acetic acid and render it useless for ethanol.

Calculating Alcohol Yield

The ethanol yield by weight of any feedstock is an important factor, not just in determining whether the particular crop or substrate you're considering is worth the effort, but as a yardstick to gauge whether your fermentation and distillation practices are up to par. The simplest method is to base your calculation on the percentage of the substrate's carbohydrate content, explained in the section "Analyzing the Feedstock." Though it's only an approximation, it's a fairly close one, and quite convenient since it works with both starch and sugar crops. One caveat: the calculation is based on complete conversion. Realistically, you can only expect near-full conversion because you probably won't get all the starches converted or all the sugars extracted. Elsewhere I used the figure of 85 percent as an achievable goal, so I'll stick with that.

So, first determine the carb content by looking at Table 6.3 for some examples, or by referring to the USDA website noted in the "Analyzing" section. Then, simply take that figure as a whole number and divide by 13.5. The result is the volume of fuel-grade alcohol you can expect per 100 pounds of substrate. To convert that to tons, simply multiply by 20. I'll use corn as an example. Dent corn has a 72 percent carbohydrate content. 72 divided by 13.5 = 5.33 gallons per 100 pounds. So one ton would be 20 times 5.33, or 106.6 gallons. Correcting for real-life circumstances, I multiply by .85, resulting in 90.6 gallons, which is very much in the ballpark for field corn ethanol.

Next, in Chapter 7, I'll get into the distillation process, and will continue in Chapter 8 with different types of stills and operating techniques.

SEVEN

The Distillation Process

How distillation works • Continuous distillation • Azeotropic distillation • Choosing a heat source • Co-generation • Using by-products

How Distillation Works

The distillation process is straightforward, but intricate in its details. Its purpose is to separate alcohol from the alcohol-and-water mixture that comprises the liquid of the fermented mash. Typically, liquid mash or "beer" contains 10 or 12 percent alcohol, so nearly 90 percent of the fluid volume — water — must be removed to obtain fuel-grade ethanol.

In the open atmosphere, water boils at 212°F and ethanol at 173°F. Alcohol has a higher vapor pressure than water, which means it takes less energy to change it from liquid into a gas. When alcohol and water are mixed, the boiling temperature varies, and falls somewhere between the boiling points of the individual components. The greater the concentration of ethanol in the mix, the lower the boiling point. Conversely, the less ethanol in the mix, the higher the boiling point. Since the distillation process involves a phase change from liquid to vapor by means of boiling, it's clear that as more ethanol is harvested from the mixture, the higher the temperature will have to be to maintain a vapor. Table 7.1 illustrates the temperatures required to boil an ethanol-water blend with the mixtures at various percentages.

Let's look at a simple example of how distillation works. If we boiled a mixture of ethanol and water in a container, more ethanol vapor would rise from the vessel than water vapor. If we should capture that vapor and condense it, the concentration of ethanol in the liquid will be greater than in the original mixture, and likewise the original mixture will contain less ethanol. Now, if we heat the condensed liquid and capture *its* vapors and

Table 7.1
Boiling Points for Alcohol/Water Mixture at 1 Bar Pressure

Percent Ethanol	Percent Water	Boiling Point, Degrees F
0	100	212.0
5	95	203.3
10	90	199.4
15	85	196.3
20	80	194.0
25	75	192.2
30	70	190.4
35	65	188.9
40	60	187.7
45	55	186.4
50	50	185.0
55	45	183.4
60	40	182.3
65	35	180.8
70	30	179.4
75	25	177.8
80	20	176.3
85	15	175.4
90	10	174.2
95	5	172.7

Note: One Bar pressure is equivalent to one Atmosphere or 14.69 psi, standard pressure at sea level.

FRANK OMILIAN

Fig. 7.1: *A transparent distillation tower is used in research to observe fractionation at the different mixture boiling points.*

condense them to liquid, the concentration of ethanol in the condensate will be greater yet.

This procedure, called rectification, can be repeated until most of the ethanol is drawn off, and in fact this is how the earliest stills raised the proof strength of the final product. But things are not that simple. An *azeotropic* condition develops when the mixture reaches around 96 percent ethanol, or 192 proof. At that concentration, the ratio of ethanol molecules to water molecules achieves a balance, and methods other than distillation must be used to further enrich the condensate. These include the use of a dehydration device known as a molecular sieve, which uses aluminum-silicon to separate water and carbon dioxide from the ethanol, or a process that introduces benzene to produce a ternary or triple azeotrope to bring the boiling point below that of the ethanol, which is then drawn off separately.

Though both these methods are practicable on a large industrial scale, benzene is no longer used to dehydrate alcohol, in part because it's a dangerous carcinogen. And the cost of using an industrial molecular sieve in a small-scale

Molecules in Motion

The evaporation of water is a good example of the relationship between heat, temperature and pressure. Water left in an open container at room temperature will evaporate slowly, because there isn't sufficient energy available to move the water molecules rapidly. When heat, or energy, is applied, those molecules get agitated and collide with each other, causing friction and expansion of the water, making it less dense and increasing its vapor pressure. When the vapor pressure reaches the same as that of the surrounding atmosphere, the water boils. Because of a phenomenon known as latent heat of vaporization, it takes an additional and significant input of energy to make that change from a liquid to a gas. (Conversely, when the process goes in the other direction — as when the vapor condenses back to a liquid — that heat energy is given up.)

If the water is held in a sealed container, it will not be able to expand freely — steam vapor takes up hundreds of times more volume than water in its liquid form in the open air. As more heat is applied, the vapor has no place to go, and pressure in the container builds up. The greater the pressure, the higher the boiling point or temperature required to get it there. When pressure is removed — or a vacuum created — the boiling point lowers. Hence, at sea level, water boils at 212°F; increasing the pressure to 15 psi (in a pressure-cooker, to use an example) makes it boil at 250°F; drawing a vacuum of 24 inches of mercury (in/Hg) drops the boiling point to 140°F. (A lower boiling point is useful for vacuum distillation, described in the following chapter, because it reduces the temperature at which boiling occurs.)

So the heat energy (or Btu's) used can remain the same while the temperature varies — all related to how densely or loosely packed the molecules happen to be within their environment. That's why cooking food at high altitudes will take longer unless a pressure cooker is used to bump up the temperature.

Table 7.2
The Effect of Vacuum on Boiling Points

Vacuum, In. Hg., Gauge	Boiling Point Water, Deg. F.	Boiling Point Ethanol, Deg. F.
29.81	21	18.0
29.39	60	48.9
27.99	100	80.1
26.47	120	96.7
25.39	130	104.9
24.03	140	114.0
22.34	150	122.2
20.26	160	130.4
17.71	170	138.7
14.63	180	146.8
6.45	200	163.2
5.46	202	164.8
4.44	204	166.4
3.39	206	168.1
2.29	208	169.7
1.17	210	171.3
0.00	212	173.1

Note: A perfect vacuum of 29.92 In. Hg. is largely theoretical for our purposes. Maximum vacuum draw in practical applications will be in the range of 25 or 26 In. Hg. Ethanol purity at 99.5 percent.

operation isn't usually worth the expense, since 99.9-percent pure alcohol isn't necessary to run engines or heating equipment. The only reason fuel alcohol is distilled to that degree is to allow it to blend with gasoline when making gasohol or E-85, an 85/15 percent mixture of ethanol and unleaded regular. Anhydrous, or water-free alcohol, has to be stored in sealed containers or tanks with special vents.

There are other, less costly methods of drying ethanol using caustic lime (calcium oxide) or rock salt (sodium chloride or calcium salts). Details about this are given later in this chapter.

Table 7.3
Alcohol/Water Equilibrium at 1 Bar Pressure

Percent Ethanol in Liquid	Percent Ethanol in Vapor	Saturation Temperature, Degrees F
0	0	212.0
1	10.3	210.1
2	19.2	208.5
3	26.3	206.9
4	32.5	204.8
5	37.7	203.4
6	42.2	202.1
7	45.9	200.9
8	48.8	199.5
9	51.1	198.5
10	52.7	197.2
20	65.6	189.2
30	71.3	184.5
40	74.6	181.7
50	77.4	179.6
60	79.4	177.8
70	82.2	176.2
80	85.8	174.3
82	86.8	174.0
84	87.7	173.7
86	88.8	173.4
88	90.0	173.2
90	91.2	173.0
92	92.6	172.9
94	94.2	172.8
96	95.9	172.7
98	97.8	172.8
100	100	173.0

Note: One Bar pressure is equivalent to one Atmosphere or 14.69 psi, standard pressure at sea level. Saturation temperature is the liquid's boiling point.

The Art and Science of Rectification

In old-time stills, the redistillation process took place in separate vessels, but with modern equipment, it's all done within a fractionating column, a large pipe sometimes referred to as a tower. The column essentially allows a series of mini-stills to be stacked one atop the other. The rectification takes place in layers, and each layer is defined by a ventilated plate.

The first column stills were overly complicated in their construction. The plates were perforated by tubes, which extended a short distance above the surface of each plate. The tubes were covered with ventilated caps that allowed vapors to pass through. Each plate also included a downcomer, a longer tube suspended from the plate above it that ended within a shallow well built onto the surface of the plate. The top of each downcomer rose slightly beyond the surface of the plate above it. One shortcoming of this design is that the mash must be free of solids to avoid clogging those caps and risers.

Subsequent designs are significantly less complicated, and utilize drilled plates with downcomers, or perforated sieve trays with

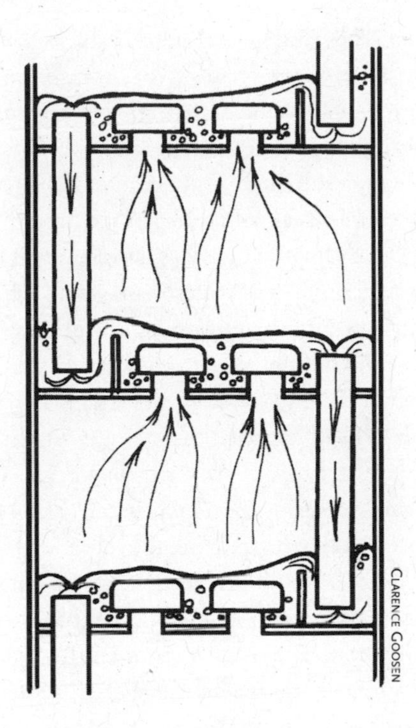

Fig. 7.2: *The bubble cap plate design.*

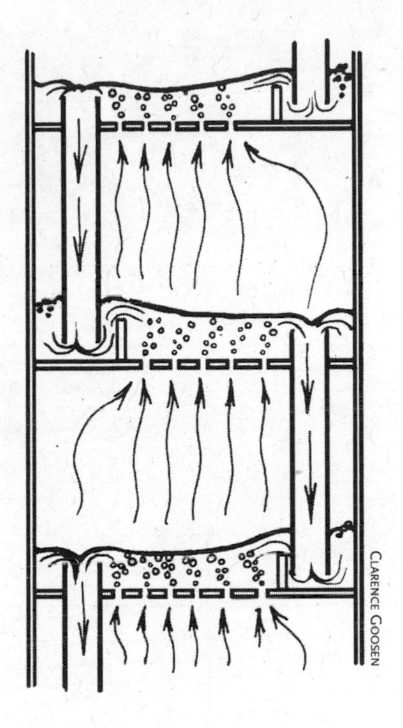

Fig. 7.3: *The perforated sieve tray design.*

integral weir dams to control flow over each plate. In all these designs, the ascending vapors do not pass through the same opening as the descending liquid, so the transfer of gas and liquid occurs on the individual plates. A specific level of liquid is maintained on each plate, and as long as the column is held in equilibrium (described below), the level will remain unchanged.

Eventually, builders developed simpler designs using perforated plates with no detailed fabrication. They are simply chambers filled with pall-ring or saddle packing shaped to provide maximum surface area for contact while still allowing up to 90 percent free air

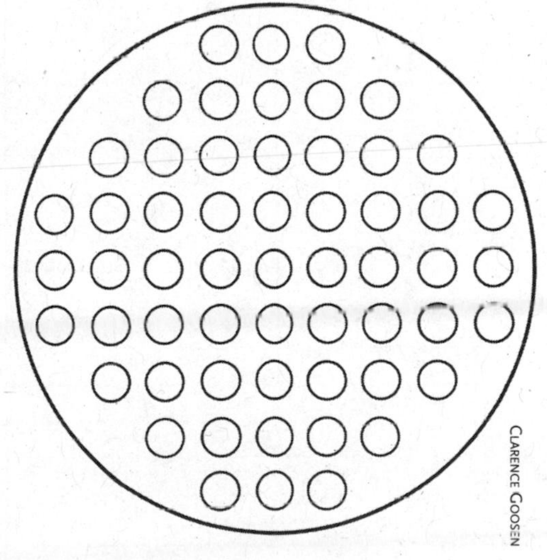

Fig. 7.4: *A stainless steel plate used with packing material.*

Fig. 7.5: *A sample of saddle packing used in a packed column.*

Table 7.4 Theoretical Plates in Ideal Equilibrium		
No. Plates	Percent Ethanol in Liquid	Percent Ethanol in Vapor
1	10.0	53.0
2	53.0	78.0
3	78.0	85.0
4	85.0	88.0
5	88.0	90.0
6	90.0	91.2
7	91.2	92.0
8	92.0	92.6
9	92.6	93.0
10	93.0	93.5
11	93.5	93.9
12	93.9	94.0
13	94.0	94.2
14	94.2	94.4
15	94.4	94.6
16	94.6	94.8
17	94.8	95.0
18	95.0	95.2
19	95.2	95.4
20	95.4	95.5
21	95.5	95.6
22	95.6	95.7
23	95.7	95.8
24	95.8	95.9
25	95.9	--

Note: Percentages reflect an initial 10 percent alcohol mash mixture. Redistillation is assumed at 100 percent efficiency, or Ideal Equilibrium. In practice, reduced efficiencies will require the use of additional plates to achieve an azeotropic condition.

space. These differential column stills allow the transfer between phases to occur right on the packing, in increments as vapors rise through the column. Packed differential columns are the simplest to build, but because of the packing, cannot deal with solids in the mash.

Depending on the still's design, it will take a certain number of plates, or stages, to achieve a high-proof ethanol product, which we can generally define as 190 proof, or 95 percent ethanol. In theory, a 10-percent mash mix would have to pass through 25 stages to achieve that proof strength. But because of normal losses of efficiency, it can take 44 or more plates to actually get the job done.[1]

There are a few things that can be done to increase the proof of the ethanol product in distillation. The simplest is to let the process work longer. It is the nature of distillation that it takes about as much time to get high-proof product from the final 25 percent of a batch run as it does to distill the first 75 percent. (A run is the period of time it takes to finish one distillation.) Because the alcohol content of the mash is so low in the final stages — around 2 percent — additional time is needed (and may be well invested) to wring as much ethanol out of it as possible. At some point, however, the energy expended offsets the benefits, and it may be better to simply let the remainder run through the column as "low wine" at 30 or 40 proof, to be collected separately and added to the next distillation batch when you're ready to make a run.

For the sake of efficiency, further methods have been developed to produce a high degree of purity without investing so much time and energy. One is to remove heat from the column through the use of heat exchangers. These are simply water-cooled coils or jackets strategically placed within the column to allow precise temperature control at critical points in the distillation process. Another method is called refluxing, which returns a portion of the condensed ethanol back to the top of the column to increase conversion.

Both methods can be used, or each individually, depending upon the still design. The control provided by these enhancements comes at the cost of initial expense and complication, but for serious ethanol production, the more sophisticated designs are all but essential to keeping costs per gallon in the range of the cost for conventional fuels.

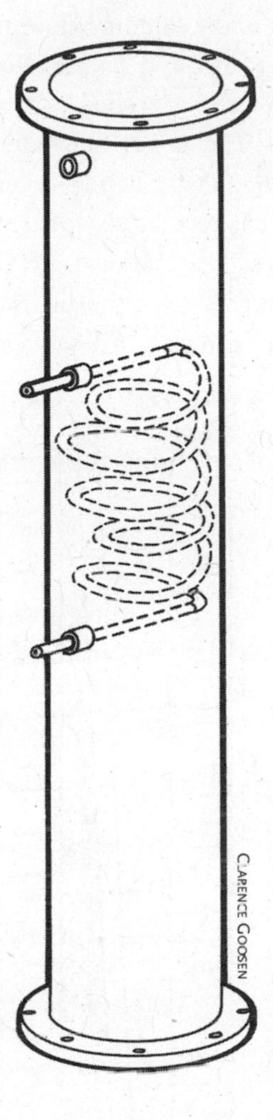

Fig. 7.6: *Water-cooled coils, or heat exchangers, remove heat from the distillation tower.*

Continuous Distillation

To keep things simple, what we've been discussing so far is batch distillation, or a single-run process where a predetermined volume of mash is prepared, heated and sent to the distillation column, after which the greatest amount of ethanol possible is retrieved from the system. Once the maximum percentage of ethanol is recovered from the mash, the system is cleaned out and the process repeated all over again.

Continuous-feed distillation automates the process and makes it flow smoothly without stopping. Continuous stills are more costly to build than batch stills, partly because of their column construction and partly because of the monitoring sensors and controls needed to make them work. Continuous distillation uses a multiple-plate fractionating column described earlier, but feeds the fermented beer into the column at its midpoint. Unlike the typical batch process, where vapors enter the column, continuous feed means liquid mash, particles and all (it is for this reason that packed columns are seldom used with continuous-feed designs unless the mash is prefiltered before it's fed in). To resolve the clogging issue while keeping costs down, some designs use perforated plates for the part of the column below the feed (the stripper section) and packing for the part above (the rectifier section), which receives only vapors. The feed is maintained at a constant rate and the distillation is consistent at a specific proof strength. The spent mash is likewise drawn off at a constant rate.

All this occurs only if the still is kept in equilibrium, the term used to describe the balance of vaporization and condensation within the column through the correct management of temperature and pressure (see the sidebar "The Principle of Equilibrium"). Equilibrium is controlled by the amount of heat introduced at the reboiler, or the lower portion of stripper column. With the still at equilibrium, each plate stays at its own steady boiling point and proof, which makes it possible to extract fusel oils and other contaminants at the points where they stop vaporizing in the column. This is a side benefit to commercial operations that need clean alcohol for medicinals or cosmetics.

Fig. 7.7: Stripper and rectifier columns are separated in a two-tower design, though they act as a single column in practice.

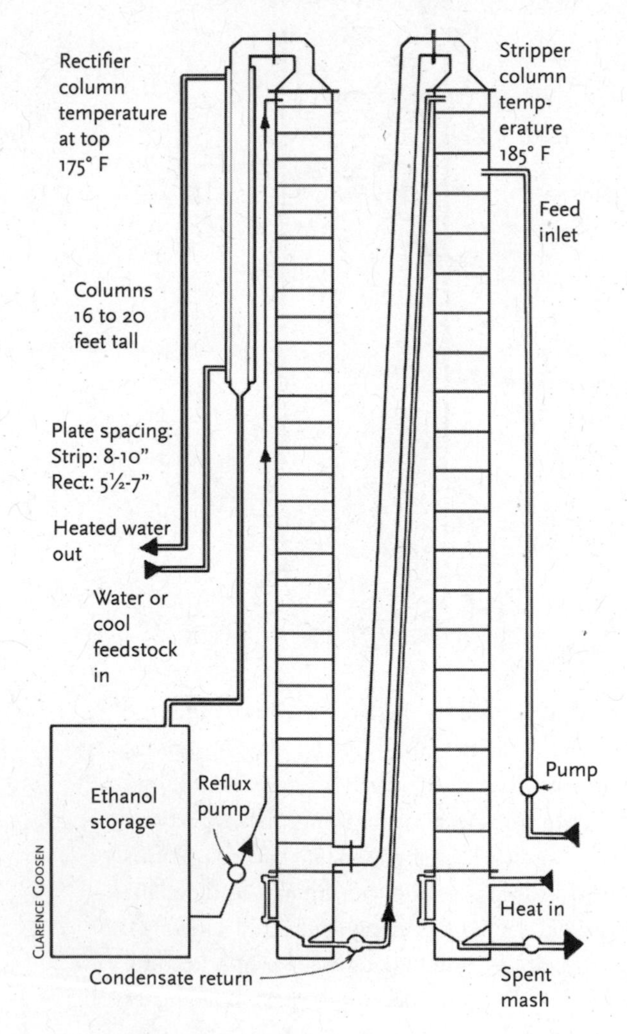

It is not at all difficult to overheat the column, especially with certain types of still designs, or if the heat source isn't consistent, as in a wood-fired boiler still. The effect is the same as removing plates or stages, because once the column is beyond the optimal temperature, fractionation and equilibrium cannot occur. Therefore, the use of heat exchangers to modify temperatures within the column is necessary to allow minute levels of control.

Continuous-feed columns are typically taller than batch-feed columns to accommodate the constant flow of material and to achieve the control needed for efficient distillation. The height differential is such that continuous-feed columns are frequently split into two columns — a stripper and a rectifier — even though they are treated as one. The vapors coming off the stripper column are piped down to the bottom of the rectifier so they can rise for further distillation and the liquid reflux descending from the rectifier is pumped to the top of the stripper so it can fall to the reboiler to be mixed and vaporized with incoming mash.

One major advantage of continuous distillation is that little heat energy goes to waste if it's recycled properly. Once the operation is running, hot water from the reboiler can be used to preheat water for mash cookers, or to preheat fermented mash in a heat exchanger before it enters the column. As you would expect, additional pumps and plumbing are needed to make this happen, which adds to the cost of the system.

Generally, a small-scale operation is better served by a batch-feed system because of its lower cost and ease of operation. "Small scale" can be loosely defined here as anywhere from a 2-inch diameter micro-column still producing less than one gallon of alcohol per hour to a 12-inch column still with an output between 25 and 35 gallons per hour. (Doubling the diameter of the column increases the still's capacity roughly by a factor of four.) When volumes greater than that are the goal, it's probably better to shift to a continuous-feed system and invest in monitors and controls rather than putting the money into larger-diameter columns and the additional cost of packing materials.

Azeotropic Distillation

Earlier, I mentioned that there is a point of diminishing returns where straight distillation of ethanol is concerned. At about 192 proof, the alcohol-water mixture forms too strong a bond to be separated by heat alone, and the mixture becomes what's known as an azeotrope. Now let me be clear on this: ethanol at that percentage is ideal for most people interested in making alcohol on a small scale.

Table 7.5
Column Diameter and Output Capacity

Column Diameter, Inches	Capacity, Gal./Hour	Area, Sq./In
2	.75	3.15
3	1.5	7.07
4	2.6	12.56
5	4.2	19.64
6	6.0	28.27
7	8.2	38.49
8	10.7	50.27
9	13.5	63.62
10	16.7	78.54
11	20.2	95.03
12	25.0	113.10

Note: Alcohol yields at 190 proof.

It's a hydrous alcohol and works quite well in vehicles, furnaces and gasoline-powered equipment such as mowers, tractors, pumps and generators. The small amount of water in the fuel poses no problem for most people in most situations. The Brazilian Flex Fuel cars I spoke of in Chapter 1 have been running on hydrous alcohol fuel for over half a decade.

Here in North America, Flex Fuel cars and trucks will likewise run fine on hydrous alcohol. There are probably only two situations in which the 4 percent or so of water in the fuel can cause problems. One is if someone lived in a climate where winter temperatures drop to –15 or –20°F range, and mixed home-brewed ethanol and gasoline to power an engine that hadn't been modified to run on alcohol. The extreme cold would cause the mixture of fuels to separate into layers, and soon enough straight alcohol would be going through a fuel system not optimized for it. At more reasonable

The Principle of Equilibrium

The purpose of a distillation column is to create fractionation, or increasingly stronger cuts of ethanol. Here's the simple explanation: As the alcohol-water vapors enter the column, they lose their heat to the surrounding materials and condense. The condensate falls down with gravity while more hot vapors continue to rise. As the condensate mixes with the ascending vapors, its water, which is less volatile, releases its heat to the alcohol, which is more volatile, and condenses. This causes the alcohol in the condensate to be revaporized, continuing a process known as enrichment. The cycle of alcohol vaporization and condensation, and subsequent stripping and descending of water, called the countercurrent stream, is what increases the concentration of alcohol to make a cut, or fraction.

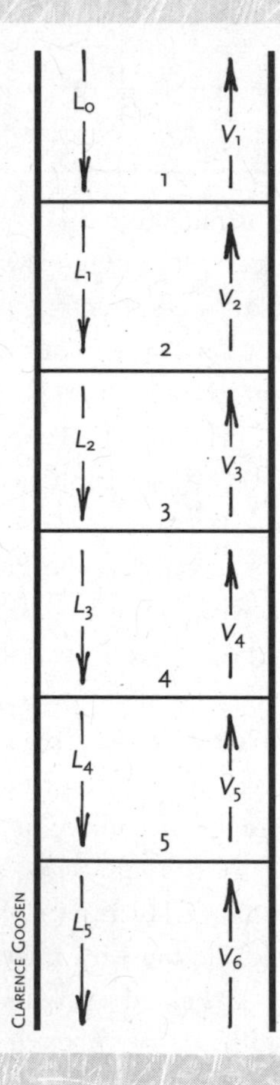

Fig. 7.8: *During fractionation, a countercurrent stream of liquid and vapor causes distillation to occur at each stage, enriching the alcohol-water mixture.*

Each plate or stage within a distillation column redistills the alcohol-water mixture to a certain degree before it passes on to the next stage. Accordingly, the concentration of alcohol in the mixture increases from one stage to the next — in other words, the proof increases. At each stage, both a liquid and a vapor stream must be present in order for fractionation to occur.

In order to keep the liquid and vapor streams flowing, temperatures must be controlled very effectively, since each stage represents a different percentage of alcohol and water and therefore a different boiling point. The process cannot move on to the next stage until it achieves the correct temperature at the existing stage. The term "equilibrium" refers to the fact that the exchange of liquid to vapor and then vapor to liquid are equal — the number of molecules that exit the liquid as vapor must be the same as the number returning as liquid from the vapor.

The efficiency of the column depends on it being in equilibrium as a whole, with fractionation occurring at all the stages. With new mash being introduced and distilled ethanol being withdrawn on a continuous basis, the liquid-vapor exchange at each phase must occur at the same rate for the system to be balanced. Temperatures are controlled through the use of heat exchangers or cooling coils that remove some of the heat from the column.

BackHome photo collection

Fig. 7.9: *The jobsite generator at this remote work site is fueled by hydrous 185-proof ethanol.*

temperatures, gasoline blends with hydrous alcohol much more readily, without causing this problem, as long as the alcohol is at least 190 proof.

The only other reason someone might want to go to the trouble of making anhydrous (water-free) fuel is to sell it as a marketable anhydrous product. Though this is certainly doable, it can be energy-intensive, even with less costly unsophisticated appropriate technology. Moreover, since 99-plus percent pure ethanol is very hygroscopic — that is, it readily absorbs moisture in the air until it reaches its familiar azeotropic point — the anhydrous alcohol has to be stored in sealed, condensation-free containers fitted with a special "conservation"-grade pressure/vacuum vent that only responds to significant changes in pressure.

Drying Ethanol

One of the traditional ways to dry ethanol is the basic process of soaking it or passing it through an adsorptive material like calcium oxide. Adsorption is the surface retention of the liquid as opposed to absorption, where it is fully taken into the bulk of another substance.

Calcium oxide, or caustic lime (CaO), adsorbs water but not alcohol. Hydrous ethanol steeped in a container of caustic lime will "slake" the lime, creating calcium hydroxide ($Ca(OH)_2$) laden with water. The calcium hydroxide precipitates to the bottom of the container while the nearly pure alcohol goes to the top. It takes about 35 to 40 pounds of lime to remove one gallon of water from the alcohol, so it's important to determine the proof of your hydrous product and to size the container appropriately — 100 gallons of 192-

proof ethanol contains four gallons of water, and there has to be room for the fuel and the lime.

After a soaking period of 24 hours, the liquid can be drawn off, distilled at 173°F, and condensed. What remains in the bottom of the distiller vat can be dried and regenerated to calcium oxide, or the liquid distilled again to recover any residual alcohol in the flakes. When temperature readings approach the boiling point of water, the alcohol content has been exhausted and the liquid can be added to the next distillation run.

Either drying or redistilling the residual material for alcohol recovery is energy-intensive. The calcium hydroxide drying process requires temperatures in the 350°F range and cannot be done with an open flame source. Dry heat delivered through a heat exchanger, or a non-combustible heated gas like carbon dioxide should be used, and the regenerated lime can't be exposed to open air or it will absorb ambient moisture. For occasional alcohol-drying jobs, it may be a better use of resources to leave the lime in its slaked state rather than trying to re-use it.

Another drying method isn't quite as effective but might be acceptable on a small-batch basis. The hydrated alcohol can be run through a container of rock salt or calcium salts, which will draw out the water as a desiccant does. A simple gravity-feed system is adequate, and the de-hydrated liquid can be collected through a fitting at the bottom of the container (it may have to be filtered to remove any residue). The salt can be dried for re-use in a hot-air bed.

Fig. 7.10: *Wood fuel makes an economical heat source for small stills, as heat can be concentrated in a firebox.*

BACKHOME PHOTO COLLECTION

Choosing a Heat Source

Living in a rural country environment has habituated me to the use of wood as a fuel. Even when natural gas, propane, fuel oil or electricity is readily available, rural residents gravitate toward wood heat for at least a portion of their needs. Wood is renewable, it's clean-burning if combusted properly in the presence of air, and it embodies a lot of energy on a per-weight basis. Less than 1½ cords of white oak provides the same energy as 11.5 gallons of fuel oil.[2]

For a small-scale alcohol producer who needs a reliable and inexpensive source of heat, wood should be given serious consideration. All the drawbacks that dull the prospects of wood for home heating — storage, inconvenience, untidiness, dry heat — are inconsequential when you're operating a firebox outdoors or in a work building. Wood is probably not the choice for an urban distillery or a location where it must be hauled from a distance. However, in many parts of the country, wood has become a "value-added" commodity that holds a cost only once it's been processed. In raw form, on the stump or as logs, it can still be relatively inexpensive. Exurban and rural communities that are experiencing a surge in development often cannot dispose of logs unsuitable for milling (which is a large percentage of the total) without paying a tipping fee at the landfill. Public works and highway expansion projects that come with growth and development are also a source of cheap or even free wood.

The key to applying wood heat properly is in designing the firebox and cooking vat or mash pot in such as way as to expose as much of the vessel's surface area to the heat source as possible. This encourages the transfer of heat to the liquid within, though much of it will still escape out the flue pipe. Building an enclosure around the fire — a firebox — is critical to thermal efficiency to prevent the wholesale loss of Btu's and to encourage the wood's complete combustion. A firebox with a brick or concrete block surround lined with firebrick or a castable refractory material (a kind of fire-resistant cement) will help to retain a large portion of the heat generated.

One of the disadvantages of wood heat is that wood is not a uniform fuel source and temperatures can sometimes be tricky to control

Table 7.6
Considerations for Heat Source Selection

Heat Source	Heat Value (dry), Btu/lb.	Form	Special Equipment	Boiler Type	Advantages	Disadvantages
Agricultural residues	3,000 - 8,000	Solid	Handling, collection	Batch burner fire tube Fluidized bed	Inexpensive, on-farm	Low bulk density requires large storage capacity
Coal	9,000 - 12,000	Solid	Smokestack scrubber	Grate fire tube Fluidized bed	Available technology	Air quality issues, cost
Wood	5,000 - 12,000	Solid	Chipper or feeder	Fluidized bed	Low cost, clean-burning	Not uniformly available
Municipal solid waste	8,000	Solid	Sorting	Fire tube Fluidized bed	Inexpensive	Limited to urban applications
Geothermal	N/A	Steam, hot water	Heat exchanger	Water tube	No fuel cost	High initial cost
Solar	N/A	Radiation	Thermal collectors	Water tube	No fuel cost	High initial cost
Wind	N/A	Kinetic	Turbine, batteries, heat sink	Electric	No fuel cost	High initial cost

Note: Adapted from *Fuel From Farms*, US Dept. of Energy

with any real accuracy. Wood's moisture content will vary if it's not dried with consistency, which changes the fuel's Btu output. With a bit of experience, you can learn to control the fire's intensity by adjusting the amount of air admitted through the damper or air inlet. Aspiration can either be natural, or preferably blower-driven for improved combustion. You can buy universal cast-iron doors with integral inlets, or fabricate your own air controls using sections of black pipe and slotted caps. Be aware that doors made from angle iron and plate steel will eventually warp in the presence of high heat, so you'll either have to protect them with refractory on the inside or use cast iron.

Solar Energy and Heat-Transfer Fluid

While I was at *The Mother Earth News,* we undertook a number of projects that used solar energy to either heat or distill fermented mash. As righteous as some of these efforts were, we eventually concluded that low-grade solar heat — energy obtained without using intricate concentrated collectors — was not intense enough to deliver the level of heat needed for direct distillation.

CLARENCE GOOSEN

Fig. 7.11: *Concentrated solar thermal energy being used to heat an oil medium for distillation in a 3-inch packed column, seen at rear.*

Early on, Minnesotan Lance Crombie (see Chapter 1) helped us launch a grass-roots "back to the land" movement of do-it-yourself alcohol producers through his early experiments with flat-plate distillation. Other efforts in that vein followed — we reported on an Alabaman's efforts to solar-distill alcohol, and we built a flat-plate distiller ourselves. After all was said and done, however, both the quantity and proof levels achieved through this method were not satisfactory for any serious production effort.

One of the more elaborate projects to come out of the solar research was Clarence Goosen's concentrating design that focused the sun's energy on an enclosed and insulated heat exchanger, and used oil as a heat-transfer medium. The hot oil was circulated with a salvaged automotive power steering pump and sent thermal energy directly to a set of heat exchanger coils built into the still's mash boiler. This enterprise succeeded because the solar energy was concentrated on one spot and could generate enough heat to overcome losses through conduction and convection. On the down side, it's a bit complicated, requires physical space, and relies on consistent sun to work effectively. A conventional gas backup system would be highly recommended if one were to use something like this on a regular basis.

Rather than simply to dismiss solar energy as a viable source of heat, it would be better to think of it as a good supplementary heat source. The 160°F to 180°F temperatures generated by an active flat-plate collector are sufficient to preheat water or provide supplementary heat to mash. Since temperature differential plays an important part in how effective the heat transfer is, using solar to bring 45°F water to a warm range is more productive than trying to heat already warm fermented mash.

Other Heat Sources

There are a variety of other energy sources that work quite well to either cook mash for fermentation or boil fermented mash for distillation. The problem is that they are generally not cheap and certainly are not renewable. Nonetheless, they shouldn't be ignored, because some individuals may have a cost-effective source of this kind of energy, and these conventional fuels can always be used for backup heat.

Industrial steam power may be available for a larger cooperative effort, especially if it's co-located with some other type of production facility. Daniel West's fruit-waste processing still described in Chapter 5 uses electric heating elements for his small-scale operation. The Painterland Farm in northern Pennsylvania, profiled in Chapter 10, has access to rice coal (a fine-grain anthracite coal), which is a cost-effective energy source in that region.

What about alcohol fuel? The short answer is that it's worth more as a motor fuel than it is as a process fuel because of the effort that goes into distilling it. For a stationary operation such as a distillery, there are better and cheaper options than alcohol. The transportable and Btu-concentrated liquid fuels in most cases are best reserved for transportation vehicles.

Another option is residual biomass from the stalks or cellulose portion of any feedstock crops you may have access to. Small and large alcohol producers in Brazil routinely burn a cane residue called bagasse, and the silage and residue from certain crops in the US can also serve as

Fig. 7.12: *At a good wind site, a homebuilt or commercial wind turbine can supplement heating requirements for a small ethanol fuel operation. New "homebrew" permanent magnet designs are very inexpensive on a per-watt basis.*

a fuel feedstock. One already in use is corn stover, but other hulls and stalks are suitable.

Renewables other than solar can also be a source of energy for those in the right location. Small-scale hydroelectric power is probably the cheapest form of electricity available once the equipment has been capitalized. Unlike wind, which can be sporadic, hydro does its thing 24 hours a day, seven days a week, and so on. Even a moderate wattage output adds up to significant power over a period of time. Small hydro operations routinely "dump" excess power into heating elements to make hot water so the energy is not wasted. While it's not quite free lunch, it is a source of very reasonably-priced electricity.

Wind energy, too, is worth considering if you happen to be in a good wind region. Site assessments for coastal and plains locales are not expensive and the cost of wind turbines has decreased on a per-watt basis from times past. With increased interest in wind, there are also some excellent options for do-it-yourself turbine projects that use low-cost permanent-magnet designs to reduce capital investment by a factor of ten or more.

Co-generation

I want to mention another source of heat energy before moving on to alcohol by-products. Co-generation is the application of waste heat from one operation to provide heat for a secondary operation. I already talked about the use of steam above, but there are other energy sources on a far smaller scale that are also fair game.

A diesel or gasoline-powered generator or pump creates a lot of heat when running. Actually, as much as one-third of the energy introduced to an engine as fuel is wasted as heat through the coolant in the radiator. This heat can be directed, very simply, to supply secondary heat to a water preheater or to a liquid-to-air heat exchanger to dry residues or calcium salts, or to provide space heat. You may not have a pump or generator on your site, but if an internal-combustion engine is any part of your operation, it shouldn't be overlooked.

Using By-products

The stillage, or leftover mash from an ethanol run, can be fed to farm animals as a protein supplement. These distiller's feeds are essentially stripped of alcohol content but retain concentrated nutrients that include proteins, vitamins, minerals, fats and yeast formed during fermentation. The removal of starch and carbohydrates in making alcohol leaves the digestible nutrients at a concentration three times stronger than they were in their original form. So, with corn, the percentage of protein is normally about 9 percent, but increases to 27 percent in the stillage left after ethanol production.

There are four types of distiller's feeds: distiller's dried grains (DDG), distiller's solubles (DS), distiller's dried grains with solubles (DDGS), and condensed distiller's solubles (CDS). The DDG are the solids separated from the mash after distillation. Solubles are the water-soluble nutrients and very fine particles that cannot be easily separated. The condensed solubles and a dried form of these are made from the liquid solubles in the stillage.

Feed can also be produced as wet distiller's grains (WDG) to save energy, but usually the solids are separated out using a screen, press or centrifuge of some type and then dried for storage. Moisture content will determine the practical length of storage time, but evaporative drying is costly, so distiller's feeds are often used soon after being processed.

Mature cattle can consume about 7 pounds of dry stillage per day; it should provide no more than 30 percent of the animal's total feed ration in a mix. Other livestock such as poultry, sheep and swine can also benefit from dried distiller's grains on a more limited basis. A local agricultural university or your county or state extension service can provide additional information on feed supplements and particular livestock needs.

Carbon dioxide is the noncombustible and colorless gas that we humans and all animals form when we exhale after breathing in oxygen. During fermentation, yeast manufacture quite a bit of carbon dioxide before they get

Table 7.7
Analysis of Distiller's Dried Grain Feeds: Corn

	Distiller's Dried Grains	Distiller's Dried Solubles	Distiller's Dried Grains w/ Solubles	Condensed Distiller's Solubles
Moisture, percent	7.5	4.5	9.0	55
Protein, percent	27.0	28.5	27.0	25.4
Fat, percent	7.6	9.0	8.0	20.0
Fiber, percent	12.8	4.0	8.5	1.4
Ash, percent	2.0	7.0	4.5	7.8
Total Digestible Nutrients, percent	83	80	82	92

Note: Sourced from Distillers Feed Research Council, Cincinnati, Ohio and *Angus Journal*, July 2007.

CLARENCE GOOSEN

Fig. 7.13: A small methane digester uses livestock manure and stillage to produce a combustible gas fuel very similar to natural gas.

on to the business of making alcohol. In most cases, on a small scale at least, this CO_2 is simply bubbled off through a fermentation lock and allowed to escape into the atmosphere. But carbon dioxide has value in many types of businesses and industries.

It's used in the bottling industry to provide the fizz in soft drinks; it's a benign food preservative and food packaging agent; it's a source of pressure for fire extinguishers, spray paint, and foams or sealants. And, because plants inhale carbon dioxide and exhale oxygen, it's valuable as a crop and greenhouse plant enrichment and growth tool. CO_2, or "greenhouse gas," is actually quite beneficial to the growing industry and is only an environmental problem when it supplants oxygen in the air we breathe.

Realistically, many small alcohol plants cannot generate enough CO_2 to justify the investment in the scrubbers, compressors and storage containers needed to market the gas on a wide scale. But if there were a greenhouse on site or a grower nearby open to an arrangement for putting scrubbed gas to use (CO_2 should be washed of trace elements from distillation), it may be worth going through this simple step to get rid of it in a useful way.

Before leaving the by-products discussion, it's important that you understand that stillage is truly a key co-product to distillation. Depending on how much effort you want to contribute to the operation, it also can be used as a fertilizer, it can be composted, and it can even be used as a feedstock in methane production, which opens up an entire world of possibilities with regard to generating energy for cookers and still reboilers. Methane constitutes about 94 percent of what we know as natural gas, so with some minimal processing (removal of caustic sulfur dioxide and subsequent drying) it makes an excellent burner fuel.

EIGHT

Preparation, Fermentation and Distillation Equipment

Cooker vats and equipment • The basic distillery • Advanced packed column still • Vacuum distillation • Operating a batch still • Alcohol storage and handling.

Your ability to make fuel alcohol will likely be more a question of management and attention to detail rather than a matter of how sophisticated your equipment is. But, from an economic standpoint, assembling the right equipment can mean the difference between making fuel at a cost that's worth your time and investment and creating an alternative to gasoline that may well be a satisfying experience but is too expensive to be practical.

Chapter 4 addresses the "macro" economics of making fuel ethanol and the "micro" economics of building equipment from scrap or salvaged parts, as opposed to searching for pre-owned distillery equipment or a used turnkey still. In this chapter, we'll look at the individual components needed to put together a working distillery.

The Small-Scale Alcohol Plant

Back in Chapter 3, I explained that the federal Alcohol and Tobacco Tax and Trade Bureau offers three Alcohol Fuel Producer's permits, based on annual production capacity. The Small AFP allows for manufacture of 10,000

Fig. 8.1: *A serious small-scale alcohol fuel operation takes some degree of investment.*

PAINTERLAND FARMS

proof gallons of ethanol per year, or around 5,200 gallons of 190-proof alcohol fuel — enough to serve the needs of about eight vehicles. In a cooperative or farm situation more fuel might be needed, so the next step up in capacity is the Medium AFP, permitting 10,000–500,000 gallons per year.

These groupings were not intended to correlate with any particular size still or its capacity — your needs will control that. In designing an ethanol plant, the size of the distillation column will essentially dictate how many gallons of alcohol you'll produce on an hourly basis, as represented in Table 7.5. Given an adequate supply of mash to keep the distillation column working properly, the column will make alcohol until the fermented supply is exhausted.

PAINTERLAND FARMS

Fig. 8.2: *An auger is used to transfer grain in this alcohol fuel plant at a Pennsylvania dairy farm.*

For the most part, I'm speaking in this book to those who have — or are planning — a single-tank batch still, in which one vessel is used for cooking, fermentation and distillation. Since it takes up to three days to complete the fermentation process, actual distillation can only occur every three or four days because the vessel is full of working mash the remainder of the time.

Feedstock Handling and Storage

No matter what your feedstock, a handling routine and some type of storage will be needed to stockpile the raw material prior to its being processed. What this ends up being depends on the nature of the stock. Traditionally, cereal grains have been the backbone of the commercial alcohol industry, and it's not likely that corn, wheat, grain sorghum (milo), rye or barley will fade from popularity, even as prices fluctuate for these grains. Sugar crops also make ideal feedstocks in regions where they are grown, though spoilage can be more of a concern because sugar crops are not as stable as starch once harvested, and there can also be some loss of sugar content in storage. Food process waste and cullage will require its own set of solutions, depending on how the feedstock is made available.

For the most part, you can look to agricultural sources for the type of equipment you'll need to handle whatever feedstock you choose. Besides the tanks (and the necessary water, warm water, spent mash and fuel storage) which you'll set up depending upon how you lay out your operation, you'll probably need a lifting auger for grains, and a slurry pump or other type of positive displacement pump that's capable of handling thicker liquids. Fractional horsepower pumps with small ports aren't cut out for this type of work. You'll need something in the 1½ to 2-horsepower range with 2-inch fittings and hoses to handle the liquid slop you're sure to encounter if you deal with fruit waste or other wet crop cull.

Grinders and Shredders

I've already mentioned grinders, shredders, hammermills and presses briefly in Chapter 6, but they're worth returning to because just about any feedstock you're apt to make use of will require some pre-processing. Moreover, some crops (grains) are more suited to milling than fruit is, which is more appropriately shredded to pulp.

For someone just getting started in a micro-sized operation, grains can be milled at a feed mill for a lot less cost than purchasing equipment up front. I wouldn't recommend this as a long-term practice because processing costs can really run up over time, but as a starter it's probably a good choice.

If you're ready to purchase equipment, consider scale first. Unless you can get a hammermill at a very low price (not likely) there's no point in buying one right away if you're not sure how serious you are about producing alcohol. A hammermill is a bulky, heavy-duty piece of equipment that uses spinning, swinging hammers to crush grain and other materials against a curved steel plate in close tolerances. For grain processing, the plate is perforated as a screen, and different screen sizes can be selected by changing the plate. A hammermill is a versatile implement, but it does not always deliver a consistent grind. Used construction and agricultural equipment dealers can be a good source of hammermills.

Shredders are more common and are available even in the economy range, suitable for weekend gardeners and homeowners. Pay particular attention to the design when scouting for shredders, because some use a drum and knife system while others (generally the larger-capacity ones) borrow from the hammermill design. Smaller shredders or choppers used for mulching leaves and branches generally have four-stroke gasoline engines in the 3.5 to 5-horsepower range. They can be built cheaply, with thin-gauge metal hoppers and economy bearings, so overall mass is a good measure of durability. As you move into the 8 to 12-horsepower range, quality improves with the use of large, sturdy hoppers and substantial shafts and bearings.

Shredders can handle a wide variety of feedstocks, both dry and liquid, and they can process a good quantity of product if they're large enough. Landscaping supply houses, used equipment vendors, and occasionally equipment rental outfits are all fair game as a source for shredders. It's a good idea to make certain that replacement parts are readily available from the manufacturer for the equipment you end up choosing.

Roller presses and hydraulic or screw presses are more or less specialized for certain types of crops, or they must be used in conjunction with shredding or chopping to extract juices. For grain crops only, there are also gristmills — some with substantial capacity — that

RICHARD FREUDENBERGER

Fig. 8.3: *A roller press squeezes the juice from sorghum plants. Older agricultural equipment like this is nearly indestructible and can still be had at a reasonable price.*

can be purchased from mail-order food supply houses or from agricultural outlets. Gristmills are very consistent in the grind they produce and would be a good choice for someone with a fixed source of grain feedstock.

DEWATERING APPARATUS

Prior to drying on racks or other means, wet distiller's grains or other feedstock substrates may need to be dewatered. There is commercial-scale dewatering equipment available, but small-scale producers would be better off economically by considering some simpler options or modifying other types of equipment for the purpose.

A hydraulic or simple screw press, such as the kind used to make apple cider, is a good starting point. By upscaling the design and using sturdier frame and basket components, it's possible to come away with a relatively inexpensive water-removing device. A standing hydraulic arbor press in the 12 to 25-ton range is a good example of the kind of appropriate technology that works well for livestock feed processing. Stainless steel washer tubs from commercial clothes washers or dryers are ready-made baskets for the clever fabricator.

Although it requires more investment, a used auger press makes an excellent dewaterer. This design uses a heavy, rotating auger mounted inside a sturdy perforated cylinder. The worm-driving force of the auger compresses the mash against a port at the rear of the cylinder, which can be adjusted to release at a predetermined pressure. Until that happens, water is pressed out of the material and is forced through the holes in the cylinder. Auger presses are used in many types of material handling and should be available from a used equipment broker.

Tanks

Determining the right size for your needs may be a matter of economics. Not all that long ago, when small dairy operations were routinely going out of business all across the country, used stainless steel tanks and vessels were readily available at very reasonable prices. Depending on their function, many, such as milk coolers, were equipped with agitators and cooling jackets.

With the passage of time, developing economies in China and India sourced a tremendous amount of scrap steel from salvage yards in the US and Canada, including plumbing, tanks, tractors and agricultural equipment, much of which was questionably "scrap" but sold by weight nonetheless.

The supply of reasonably priced large agricultural and industrial handling equipment has subsequently dried up, but much more recently, global economic slowdowns have pulled the bottom out of the scrap steel market, so salvage tank prices should be shifting to a buyer's market, even at the individual level.

Size should actually be determined by how much volume you're prepared to distill at one time. For a 6- or 8-inch column, appropriate for running a batch economically in a time frame between 4½ to 8 hours, a 750 to 1,500-gallon vessel will be sufficient. For the cost-conscious, a 6-inch column mounted on a 500-gallon tank is an ideal starter setup in terms of initial investment and manageability.

The capacity of various sized tanks is illustrated in Table 8.1, which shows ranges from just under 750 gallons to over 5,000. Mild steel tanks are acceptable given reasonably thick wall material. Stainless steel is excellent for its longevity, but stainless is usually costly and requires special skills and welding equipment to fabricate modifications. If you intend to use a single tank for cooking, fermenting and distillation, it's important to plan for some additional capacity — perhaps 20 percent or so — for expansion and foaming in the cooking and fermenting stages.

Some consideration should be given to the vessel's configuration with regard to heating, as well. The more surface area exposed to flame, the more efficiently and quickly its contents will heat. Hence, for a wood-fired setup, a horizontal tank design would be more desirable. An oil- or gas-fired system can concentrate

Table 8.1
Vertical Tank Capacities, US Gallons

Depth, ft.	Diameter, ft.						
	2	3	4	5	6	7	8
1	24	53	94	147	212	288	376
2	47	105	188	293	423	575	752
3	70	158	284	440	634	863	1128
4	94	211	376	587	846	1151	1504
5	117	264	470	734	1058	1439	1880
6	141	317	564	881	1269	1727	2256
7	164	370	658	1028	1481	2015	2632
8	188	423	752	1175	1692	2303	3008
9	211	475	846	1322	1904	2591	3384
10	235	528	940	1469	2115	2879	3760
11	258	581	1034	1616	2327	3167	4136
12	282	634	1128	1763	2538	3455	4512
13	305	687	1222	1909	2750	3742	4888
14	329	740	1316	2056	2961	4030	5264
15	352	793	1410	2203	3173	4318	5640

Fig. 8.4: *A horizontal tank works well when used with a full-length firebox, built beneath the vessel. The flue at rear, behind the distillation column, allows combusted gasses to exit.*

heat more effectively, so a vertical tank may work for fuel oil or gas-heated systems. Larger distilleries often use steam as a heat source, and it is injected directly into the mash or routed through strategically placed tubing, so the tank's configuration isn't as important.

Another practical consideration is access to the inside of the tank for cleanout and maintenance. If possible, choose a vessel with a large enough access hatch to allow human entry, or at least get one that will allow you to fabricate or install a purchased hatch head. A 16 by 16-inch square or 20-inch round head should be adequate, and in any case the lid should be equipped with a positive latch, and it should be well sealed. For a cooker especially, it is very difficult to clean the inside properly without physically scraping any burned residue from the surfaces. If you're cooking in a separate vat and pumping mash to the tank for fermenting and distillation, the human-sized access may not be as critical. Remember, too, that you may

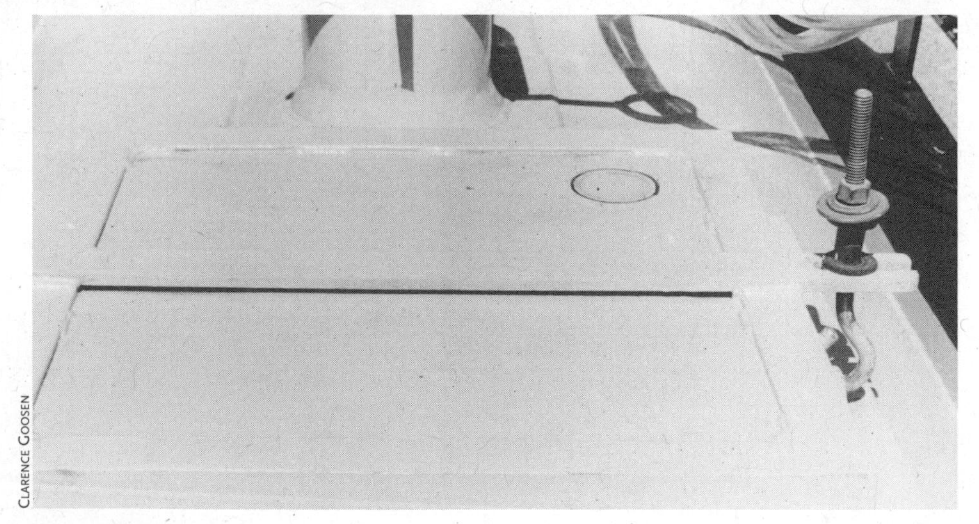

Fig. 8.5: *A large sealed access hatch with a sturdy locking mechanism helps significantly with cleanout and maintenance.*

have to modify the tank for your own plumbing needs — for example you might want to add internal heat exchangers, enlarge plumbing and drain ports to suit your transfer pumps, or include modular agitators for mixing.

Heat Sources

The subject of heat for the boiler is addressed several times in this chapter and elsewhere in the book, and it's a very important consideration to the economic viability of your operation. Raw Btu's account for a large portion of your alcohol's cost, and the more you can reduce this energy input the more profitable your venture will be.

For many, sustainable and renewable heat sources will be desirable, and these fuels fall into two categories. The first includes wood and other plant or crop biomass; these are simple and proven and are an excellent means of direct firing a boiler or cooking tank. Equipment is comparatively inexpensive, or can be shop-fabricated to suit individual needs. Many small and farm-scale operations, particularly in rural developing economies, rely on this type of very appropriate technology with little problem. The second category covers the higher-tech renewables such as solar, wind and possibly microhydro power. As you'll see later on in this chapter, these kinds of renewables can aid in producing heat, but they require substantial equipment investment to do more than supplementary service. Preheating water for cooking and heat processing are viable goals; solar energy in particular has added potential for vacuum distillation, which we'll discuss further on.

BACKHOME PHOTO COLLECTION

Fig. 8.6: *Co-generation may be an option for some ethanol fuel producers. Here, gasified wood chips fuel a 4-cylinder engine coupled to a 15-kW generator. The engine's coolant preheats water through a heat exchanger.*

Heat from co-generation may also be an option for some. An internal combustion engine set up for some other function dissipates enough energy in its cooling system and exhaust to make a noticeable dent in a small plant's energy budget. Means to recapture this lost energy through liquid-to-air or air-to-air heat exchangers are readily available, and are the stuff of home-grown dreams.

The use of hot oil as a heat transfer medium has always been in the picture, and it was one of the more promising energy schemes we used at the Mother Earth News research center for reasons you'll see in a moment. Hydraulic oil, or some type of transfer oil with a high flash point is direct-fired in a heat coil or heated in a flue, then pumped to a superinsulated storage tank or used directly to heat a boiler or cooker by means of a heat exchanger. The oil does not boil and expand as water does, so the issues of steam and pressure become moot points. Using a heat-resistant pump and plumbing, hot oil can be moved at 300-plus °F from a remote source and directed through compact heat-exchange coils mounted in any vessel requiring heat. A pressure-release valve and an atmosphere-vented expansion tank serve as simple safety controls. Heat management is achieved by using a thermostatically controlled bypass valve to shunt excess heated oil back to the heat source. A sensor in the heat generating coils also monitors temperature and modulates the fire if needed.

This brings up steam as a heat source, which is an efficient method of delivering energy, but realistically is limited to those with professional certificates. Many references to small- and medium-scale distillation invoke steam as a panacea for all energy woes, but the fact of the matter is that working with boilers, and plumbing, welding and handling live vapors at the temperatures we're talking about is a potentially hazardous practice best left to those who are trained for it. Despite steam's efficiency factor, for the most part it is generated by some form of fossil fuel, and its use requires licensed installers, periodic state inspections, and more than modest investment.

Petroleum-based energy such as natural gas, propane, fuel oil and coal — and I'll include utility-provided electric here as well, despite some nuclear and hydroelectric production — is very convenient but also very expensive. For a small experimental-type operation, an electric heating element may be a suitable choice, but economies of scale don't necessarily prove out. The biggest benefit of these heat sources is the finite temperature control they offer, especially with automated equipment. In some regions of the country, certain non-renewables are attractively priced, but this doesn't hold true everywhere.

You might ask why alcohol isn't a prime heating fuel for distillation. The answer is in its heating value, which at 76,500 Btu per gallon is less than 60 percent of fuel oil's 135,000. In Chapter 9, I'll get into why this isn't so critical an issue for engine fuels, but for now we'll leave alcohol off the list as a heat source.

Pumps and Plumbing

There is little so discouraging as having to handle large volumes of warm, fermenting liquids with 5-gallon pails in an operation that depends on maintaining a schedule. So, pumps and the corresponding plumbing that go with them are a very important part of your plan. As with most tools and equipment, you can easily get carried away with purchases, so unless you're working with a liberal budget, you can plan on pumps doing double-duty where possible and using flexible hoses with quick-disconnect fittings in addition to hard plumbing. If you plan carefully, you'll probably be able to use gate valves and iron pipe to isolate one delivery or return zone and initiate another using the same pump. Fittings and valves can be expensive in larger sizes, though, so work out a comparison between the cost of additional pumps and the cost of hardware before you jump in with both feet.

Though positive displacement pumps are generally acceptable for distillers, the gear type does not work well with mash slurry because the particles become bound in the gear teeth, damaging them or the motor. Rotary screw pumps handle slurry far more effectively. Centrifugal pumps are a good choice, but have the disadvantage of having to be "wet" to start effectively (they're difficult to prime if they're not located at the lowest point in the plumbing) and they may have problems moving thick liquids unless they have a large capacity.

Another type of pump worth considering is the diaphragm style, which uses a tough, flexible skin to expand and contract the pump chamber, thus drawing in, and then pushing out, the material to be transferred. These pumps are sturdy, nearly clog-free, and their design is such that they deliver a consistent volume.

With any choice you make, be sure the equipment is able to handle hot liquids and the delivery rate you require. Very small trial plants may get by with modest equipment, but serious operations will need to be moving 30 or more gallons per minute at 40 psi, at heights of 20 feet or so.

When laying out your operation, remember that your pumps will be handling everything from water to liquid slurry to nearly pure alcohol. Potentially, you'll need pumps for feedstock delivery, mash cooking agitation, mash transfer to fermenting and distillation tanks, condenser and heat exchanger water delivery and return, mash feed and stripper column or stillage return (in a continuous design), rectifier condensate return (in a continuous design), reflux delivery (in a direct-reflux design), alcohol storage, and cleaning-disinfecting solution, for which you'll want a separate tank as well. If you use hot oil as a transfer medium, you'll also need a suitable pump for that.

The decision to use rigid pipe or flexible hose may come down to a matter of cost, but keep in mind that it's far easier to remove and clean — or clear out — flexible hose than it is solid threaded pipe. Whatever you choose, it will have to be cleansed of bacteria on a regular basis to prevent infection of the entire system it's connected to. Speaking of connections, threaded fittings may not remain leak-free if conventional pipe sealants are used, especially once your ferment is distilled. The most available and economical choice here is to use Teflon tape and Teflon pipe compound, as it stands up well to the effects of ethanol in varying strengths.

Accessories

There is a lot that goes into setting up even a micro-sized fuel distilling operation, and as capacity increases and plans get more involved, yet more bells and whistles get added to aid in automating the system. For our purposes, I'll try to limit the accessories to the things you'll likely need for a small fuel distillery.

A set of scales is helpful for weighing feedstock to gauge how much water to add when preparing mash. Agricultural feed mixers sometimes include built-in scales, but any larger-capacity scale will suffice. It does not have to be digital or accurate to fractions of an ounce.

Probe-type dial thermometers in the temperature range you'll be using are a must. They can be threaded into fittings built into your tanks and column. For locations difficult to see into, some type of remote thermometer will have to be used. This includes a sensor fitted to the column or heat exchanger, with a cable leading to a conveniently placed dial or digital monitor.

Fig. 8.7: *An analog thermometer with probe is often the least costly option for temperature monitoring. If access is a problem, a remote thermistor with digital readout can be used.*

Fig. 8.8: *A cooker vat for a 6-inch column still. Liquid mash is agitated and transferred through the pipe at bottom.*

A pressure gauge can be helpful in determining the column's internal pressure, which correlates to the amount of heat fed into the boiler or reboiler. Remember, this is vapor pressure, so a pounds per square inch scale is far too great; a dial readout in ounces per square inch or in water column pressure is what you want.

You should have a safety pressure valve on the boiler tank, especially if you get into vacuum distillation (safety pressure valves are detailed later in this chapter). A pressure limit of 15 psi offers adequate protection. Note that the standard temperature-and-pressure valves used in domestic hot water systems have parameters different from what's really needed for a still, so they will not function correctly. For vacuum distillation you'll also need an accurate vacuum gauge (0-30 In. of Hg) and a simple vacuum breaker for the mash feed pipe in the event that the line becomes clogged (it will allow atmospheric air into the system if vacuum levels go past a predetermined point).

Flow control valves are expensive, but can be indispensable for large-column stills. Installed at the top of the distillation column where the vapors exit to the condenser, this valve constantly reads the vapor temperature and makes incremental adjustments to water flow through the heat exchanger control coil at the top of the column. Strict temperature control at the top of the column is critical in maintaining a high-proof distillate.

The Cooker Vat

The mash cooker can be a separate vessel, or, in the interest of economy, it can double as a fermenting vat and even serve as the distillation boiler tank as well. This piece of equipment is as important as the column itself, because thorough cooking is the key to successful starch-to-sugar conversion. If conversion is inadequate, a significant portion of the feedstock is wasted to inefficiency, and the cost of producing alcohol rises considerably.

Paddle Agitation

One of the simplest cookers available is a home-built design based on a standard 175-gallon fuel oil tank, a 60-inch horizontal configuration with 27-by 44-inch oval end caps. It can be direct-heated with wood or gas and uses a paddle-type agitator rotating at 40 rpm to mix the entire volume of mash with its twenty 6½-by 6½-inch 16-gauge steel paddles. The paddles turn on a 1-by 64-inch steel shaft set in bearing housings.

Slow agitation is a low-tech, low-maintenance technique that offers excellent heat and enzyme distribution and moderate aeration. A ½-horsepower gearmotor or other low-rpm motor is coupled with a chain drive to sprockets sized to achieve the 40-rpm target. The tank is fitted with a square access hatch at the top and is mounted in an angle-iron frame with a firebox below. A 6-inch exhaust flue exits from the rear of the firebox.

Pump Agitation

Agitation by pump is a more costly method, but it has the added benefit of reducing irregular-sized grains to a uniform consistency to

Fig. 8.9a and 8.9b: *End-view and side-view details of the paddles within a paddle-agitation vat fabricated from a 175-gallon or a 270-gallon fuel oil tank.*

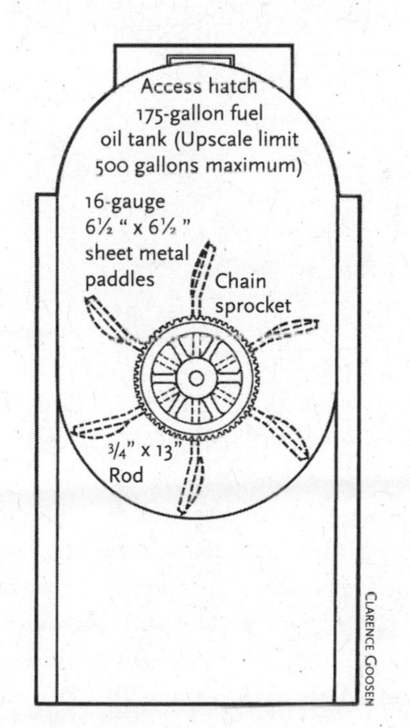

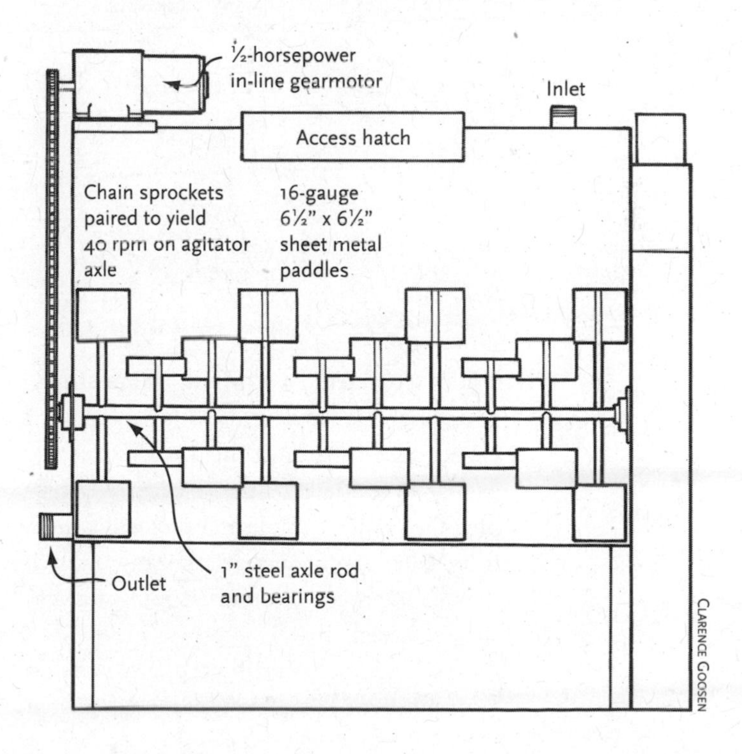

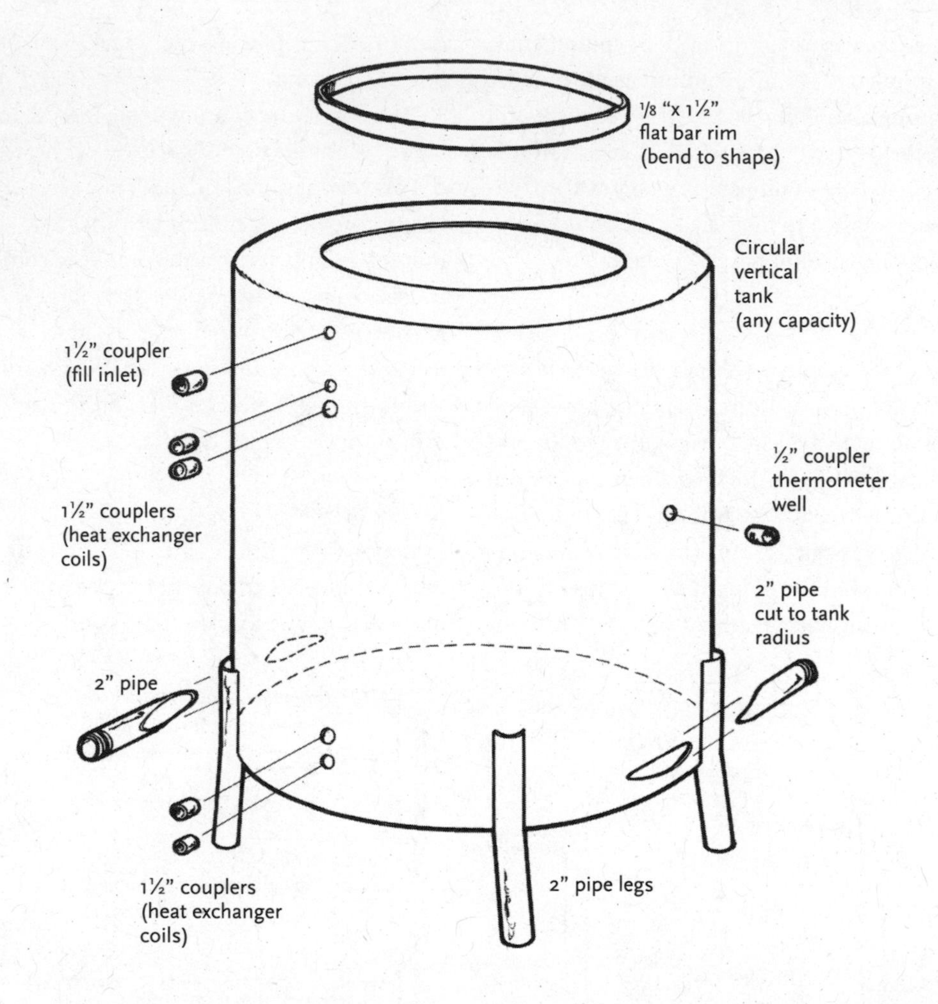

Fig. 8.10: A pump agitation cooker using a vertical tank. A vessel from several hundred to over 1,000 gallons capacity can be used.

aid in complete hydrolysis. Hammermill-ground grains result in a wide variation of particle size. Sometimes the particles are large enough to gelatinize on the outer surface, trapping the granules inside and protecting them from water and further gelatinization. Starch that is not broken down into simple sugars is simply wasted.

A self-priming centrifugal pump, or better yet, a positive displacement pump rated for at least 2 horsepower will be needed for agitation. Pumps suited for this kind of material handling can run between $300 and $800. A centrifugal pump uses an internal impeller to fling the liquid charge outward against the housing and through the outlet port. These are the least expensive of the transfer pumps but must always be charged with liquid in order to pump it. A positive displacement pump uses a diaphragm, a piston or a spiral rotor to actually push the

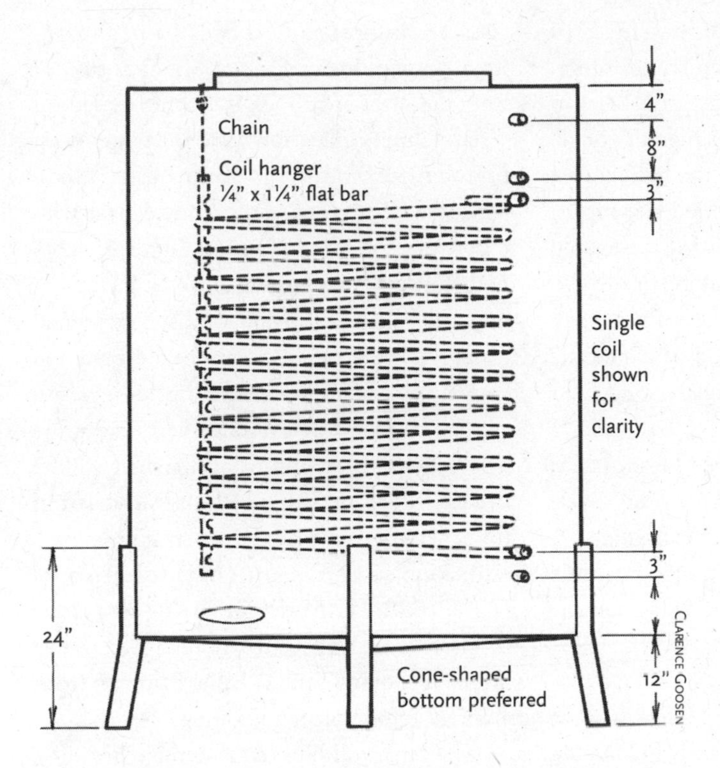

Fig. 8.11: *Heating or cooling exchange coils within the pump agitation vertical cooker tank. The vessel can serve as a fermentation vessel as well.*

Fig. 8.12: *A plumbing and flow diagram for the pump agitation cooker tank.*

liquid through the chamber. Positive displacement pumps are more costly, but they are also more consistent in their flow rate and less likely to clog or present problems with thick mash solutions.

Pumps for agitation should have inlet and outlet ports at least 2 inches in diameter. For safety, any pump used to handle alcohol should be rated to handle fuel, and would therefore be denoted as "explosion proof." Even though you may be only handling mash, the pump will be working in the presence of alcohol fumes and at some point may be used to transfer liquid ethanol, so it's best to start off on the right foot. You should also check the manufacturer's specifications to make sure all materials in the pump are compatible with what you'll be exposing it to, including heat, ethyl alcohol, particulates, and so forth.

The tank in a pump-agitated system can be any size and can serve as a cooker and a fermenter. Rather than build on a 1,000-gallon platform, it may be cheaper and more expedient to set up two 500-gallon cookers to cook two batches in a day and ferment in both vessels. Mild steel is a common material for cylindrical tanks, but the insides must be painted with heat-resistant epoxy to inhibit corrosion. Stainless will outlast mild steel by far, but cost, even as salvage, is high. If you do not plan to use direct heat on them, fiberglass and medium-density polyethylene plastic tanks may work well as fermenting vessels, but they shouldn't be used for cooking or distillation. Fiberglass is more tolerant of heat than is plastic, which will deteriorate over time when exposed to temperatures beyond their rated levels. You can expect to pay around $0.75 per gallon capacity for new plastic tanks; fiberglass will run about $1.25 per gallon. New mild steel might cost $1.95 per gallon for 12-gauge steel, but pre-owned tanks in usable shape can be had for as little as $0.50 per gallon capacity.

If you have a choice, try to get a tank with a conical bottom for drainage. Otherwise, you may have to pitch the tank slightly to one side to clear it of liquid after cleaning. The particular design shown in figure 8.11 uses two coils of 1-inch copper tubing suspended on hangers from the inside of the tank and plumbed through pipe couplings welded through the tank walls — one for heating, the other for cooling. Hydraulic oil is used as the heat transfer medium because, unlike water, it can be heated to 300 or 400°F without boiling and creating pressure. A transfer oil with a flash point over 400°F allows for a margin of safety. There are a number of options for hot-oil generators, from commercial units at one end of the scale to the concentrating solar-thermal design we researched at Mother Earth. One of the most cost-effective methods is to pump transfer oil through heat coils set in a wood-fired heating flue or firebox. Stainless steel or iron, rather than copper, is recommended for direct-flame applications, and the pump must be rated for high-temperature use. Direct-mounted oil-furnace pumps, single-stage with a flow of 3 or 4 gallons per minute, are economical and work reasonably well for this. A pressure-release valve and atmosphere-vented expansion tank must be used in the system in case the pump stops functioning.

The capacity of the tank will determine how much tubing and what diameter will be needed to deliver adequate heat transfer from the generator to the cooker. As an approximate estimation, you can figure on at least 20 square feet of coil surface area for every 100 gallons of mash to be heated. It will take four to five hours of cooking time to bring the mash from ambient temperature to boiling.

The tank will have considerable heat loss from the sidewalls and the top surface, especially in an unheated environment. To save energy, it's important to insulate the tank with closed-cell foam sheets, fiberglass batts, or any high-efficiency insulation than can withstand heat over 180°F. The exterior surface of the insulation material can be covered with 26-gauge sheet metal snap-riveted at the seams (the jacket could also be steel-strapped or duct-taped).

Distillation Columns

The distillation column can be made from a variety of materials, as long as it can be effectively sealed and is able to withstand temperatures in the range of boiling water vapors. Commercial

Fig. 8.13: *Distillation columns used in commercial research for a major beverage distillery.*

Fig. 8.14: *Small experimental stills can use polyvinylchloride pipe (PVC) for column material if temperatures are not allowed to get too high.*

columns are commonly made of stainless steel, but mild steel pipe or thinwall tubing is more readily accessible, easy to work with, and available in many standard diameters. Small experimental-type stills can even use 3-inch polyvinyl chloride (PVC) pipe for column material.

Column diameter determines the production capacity of the still, measured on a per-hour basis. Conventional practice is to match the size of the column to the tank capacity, based on how long you want each run to extend, in hours. A comfortable run would be the equivalent of a regular work shift, somewhere between six and ten hours. This isn't necessarily carved in stone, but there's little reason to vary the accepted formula. Too small a column on a large-capacity tank makes for an overly long run with the potential for extended fuel consumption. A large column on a small boiler makes it difficult to achieve equilibrium and deliver the volume of vapor needed to maintain it.

Doubling the size of the column increases its capacity by a factor of four. Table 7.5 in the previous chapter on distillation illustrates the relationship between diameter and capacity, and total area, which can be used to calculate

Table 8.2 Volume of Pipes and Cylinders		
Diameter, In.	Volume, per Foot of Length Cubic Feet	US Gallons
3	.0491	.3672
4	.0873	.6528
5	.1364	1.020
6	.1963	1.469
7	.2673	1.999
8	.3491	2.611
9	.4418	3.305
10	.5454	4.080
11	.6600	4.937
12	.7854	5.875
13	.9218	6.895
14	1.069	7.997
15	1.227	9.180
16	1.396	10.44
17	1.576	11.79
18	1.767	13.22
19	1.969	14.73
20	2.182	16.32
21	2.405	17.99
22	2.640	19.75
23	2.885	21.58
24	3.142	23.50

the amount of packing needed. Table 8.2 indicates the contents of various sized pipes and cylinders as well. If you're using salvaged or homemade materials, the cost of a larger-diameter column may be worth the investment in terms of output, but if you're purchasing commercial materials — especially packing — it may get expensive. The next section will discuss packing specifically, but no salvaged substitute can quite duplicate the performance of well-designed commercial packing.

While we're on the subject of performance, now is a good time to discuss column height. The height of the column has a direct influence on proof yield of the distillate, because the greater the height, the more water can be stripped from the alcohol/water vapor mix. There is a point, at around 192 proof, where no more water can be removed by distillation no matter how tall the column. But, by means of meticulous heat control and use of the right packing materials, column height can be reduced slightly while still maintaining a high yield. The rule of thumb is a height-to-diameter ratio of 24:1 for a 190-proof product. In other words, for every inch of diameter

Table 8.3
Design Criteria for a Packed Column

Column Diameter, In.	Height for 180 proof, Ft.	Height for 190 proof, Ft.	Pall Ring Sz. plastic, In.	Rating	Pall Ring Sz. metal, In.	Rating	Saddle ceramic, In.	Rating	Saddle metal, Mm.	Rating	Saddle plastic, In.	Rating
6	10	10-12	⅝	GD	⅝	GD	½	FR	--	--	--	--
8	10	10-12	⅝	GD	⅝	GD	¾	FR	--	--	1	EX
10	12	12-14	1	GD	1	GD	1	FR	25	EX	1	EX
12	12	12-14	1	GD	1	GD	1	FR	25	EX	1	EX
15	12	12-14	1	GD	1	GD	1	FR	25	EX	1	EX

Rating Key: EX (Excellent) GD (Good) FR (Fair)

there should be 24 inches of height; a 6-inch column would therefore be 12 feet tall, and a 12-inch column 24 feet in height. Table 8.3 shows some design criteria for batch-packed columns between 6 and 15 inches in diameter with bottom vapor entry.

Modifications to the conventional bottom-entry design were developed by my colleague, Clarence Goosen, in his early ethanol research, based on his contention that continuous-feed principles could simply be applied to batch-fed packed-column designs. His arrangement included a boiler-immersed reboiler fitted to the lower portion of the column, and a separate column fitted between the boiler tank and the center of the column, designed to introduce soluble-free vapors at the column midpoint. By adding a heat exchanger at that point, the column's capacity was improved by nearly 50 percent in some tests, though energy efficiency was somewhat reduced. Table 8.4 demonstrates diameter-to-height criteria for five different sized midpoint-entry columns in both the stripper and rectifier sections. These dimensions do not include the additional height of the heat exchanger at the top of the column, which in essence functions as a series of extra plates and brings the ratio closer to the 24:1 standard. The diameter of the introductory column can be experimented with to suit pressure and packing characteristics, but that conduit should not be so small that it restricts vapor flow. A diameter of one-half to two-thirds that of the column is a good working range.

Small-scale producers will find that insulating the column (and the introductory conduit, if used) helps to conserve heat and

Table 8.4
Diameter to Height Criteria

Column Diameter, In.	Height of Stripper, Ft.	Height of Rectifier, Ft..
6	5	10
8	5	10
10	6-8	10-12
12	6-8	10-12
15	6-8	10-12

Note: Criteria established for 190-proof distillation. The addition of a controlled upper heat exchanger will reduce the height requirements of the rectifier column by 10 to 15 percent.

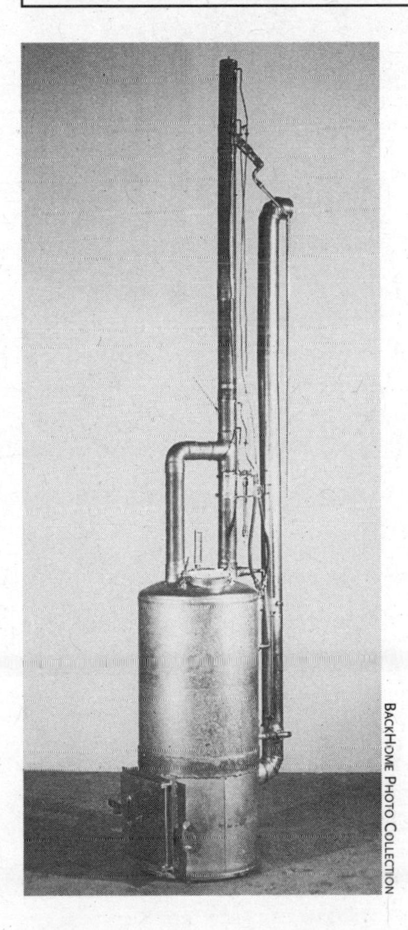

Fig. 8.15: *A midpoint-entry column design, developed by Clarence Goosen, was used in several alcohol fuel distillation designs.*

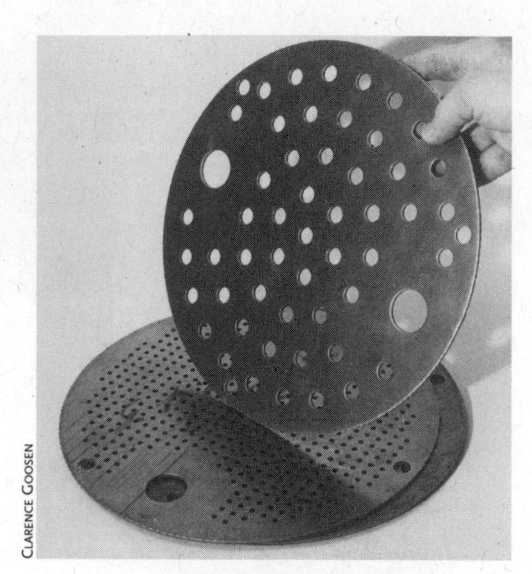

Fig. 8.16: *Perforated plates used in equilibrium-stage column designs utilize perforations, weir dams and downcomer tubes to maintain a specific liquid level at each stage.*

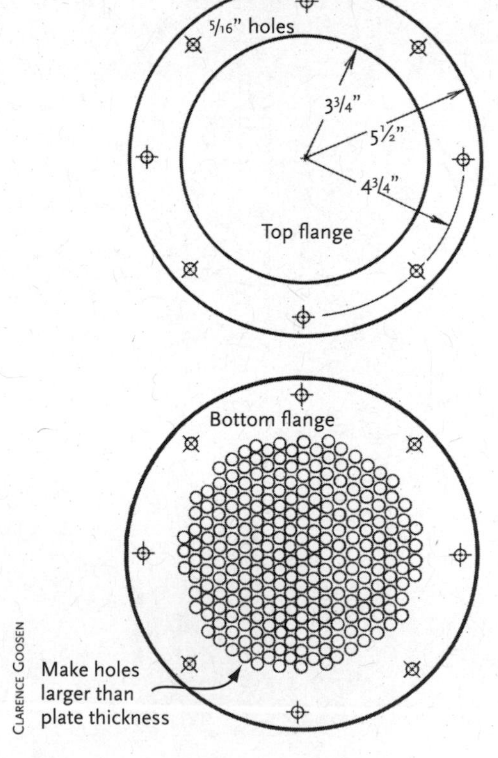

Fig. 8.17: *A flange and perforated plate set for a packed-column still.*

control fluctuations in temperature. Even a modest layer of fiberglass or a closed-cell foam jacket provides a welcome degree of consistency. Be sure to choose materials that can withstand the temperatures present at the outer surface of the column. The insulation can be secured with wire or cable ties; nothing more permanent is needed.

Plate Column Design

There is much to be said for experimentation, and in the field of small-scale distillation there has been a bounty. Practices have been borrowed from moonshine producers and concepts adapted from industrial distillers to get the highest yield from the least amount of expenditure. On an investment-versus-performance scale, I'm convinced that nothing beats a packed column for small-scale distillation.

Having said that, in moving from batch distillation to continuous distillation, there are good arguments for moving toward the equilibrium-stage designs, and those with perforated plates in particular. A perforated plate is economical to construct, as each plate contains about 8 percent free area in the form of holes or perforations. Simple tube downcomers and weir dams control the amount of liquid on the plate, which is held from draining through the perforations by pressure in the column. Plate designs are suitable for larger-diameter stills (12 inch is common), though the higher operating pressures can force heat to migrate up the column if not properly controlled, reducing proof levels.

The round plates can be precisely cut to your dimensions on a plasma or CRC machine at a local fabrication shop. Hole size is variable, but the openings should not be smaller than the thickness of the plate. Table 8.5 illustrates the range of hole diameters and their relationship to the total area of the plate and number of perforations in each. Larger holes are better at self-cleaning, but smaller openings provide superior vapor–liquid contact for rectification. Continuous stills use larger holes in the lower stripper section and smaller perforations in the upper rectifier section to mitigate mash clogging issues. The perforations should be evenly distributed in a triangular or square pitch pattern.

Downcomer tubes are sized for maximum liquid load. A 12-inch diameter stripper section can use 2-inch diameter tubes with 3-inch diameter seal caps mounted over the lower ends to prevent vapors from freely ascending the column. The rectifier section then uses 1½-inch tubes with 2¼-inch seal cups. Seal cup depth is 1¾-inches, and the downcomer tubes should extend into the cup body by at least ½-inch.

Distance between the plates is determined by how much foaming is expected from the liquid. With conventional corn and non-foaming mashes, you can get by with 8 or 10 inches of separation in a 12-inch stripper column. For wheat or other foaming grain mash, more space might be needed. The rectifier column doesn't have to deal with solubles, so foam isn't much of an issue and spacing can be reduced to 5½ to 7 inches.

Table 8.5
Hole Criteria for a 12-inch Perforated Plate

Diameter, In.	Area, Sq. In.	Number of Holes
1/16	.0031	2,918
3/32	.0069	1,311
1/8	.0123	735
5/32	.0192	471
3/16	.0276	327
7/32	.0376	240
1/4	.0491	184
9/32	.0621	145
5/16	.0767	118
11/32	.0928	97
3/8	.1105	81
13/32	.1296	70
7/16	.1503	60
15/32	.1726	52
1/2	.1964	46

Note: Thickness of plate has a bearing on the hole size. Hole diameter should not be smaller than plate thickness. Smaller holes provide superior vapor-liquid (mass transfer) contact, but larger holes in the stripper section will thwart material clogging with pulpy and particulate mashes. Plates should have approximately 8 percent free area. Stripper places can be spaced 8 to 10 inches apart, rectifier plates 5½ to 7 inches apart.

Packing

Packing is a convenient alternative to bubble-cap and perforated plates because there is no fabrication involved, yet it is still a highly effective medium for distillation. The term *packing* refers to any material used within the distillation column to create a turbulent vapor flow while diffusing liquid flow. Packing can take the form of pall rings, Raschig rings, saddles or any other design that meets the criteria.

Keep in mind that enrichment, or the removal of water from alcohol in the vapor mix, occurs when the vapors condense and revaporize. Each time this happens, some alcohol vapors rise, and condensed water particles drop into the permeable packing below, where they'll find another hot spot and revaporize again, giving up a bit more alcohol. The water eventually finds its way to the bottom of the column and back into the boiler tank, where it will be vaporized again. Since our goal is simply to produce high-proof fuel (not to simultaneously draw off by-products at distinct points on the column, as with plate stills), the packing is an efficient and very cost-effective means of reaching that goal.

Packing has to meet a number of requirements to be effective. First, it must allow a high percentage of free gas space. More voids equal higher capacity. A good design will yield 63 to 95 percent free gas space. Next, it has to offer a high surface area per unit volume to encourage condensation and ultimately reduce the height of the column. Third, the shape of each individual piece must be irregular to make a random stack and prevent a condition known as "pattern packing," in which the gas and liquid channel up and down the column without interacting sufficiently with the materials. Fourth, it should offer low flow resistance to conserve heat energy. Resistance to flow is called pressure drop, and high pressure drop, or backpressure, requires more heat to drive the vapors and causes flooding. Too-low pressure drop rushes the vapors through the column and results in low proof. A fifth desirable characteristic is shape, specifically one that creates turbulent contact between phases while evenly dispersing condensed liquid ... but not at the expense of excessive backpressure. As if all this is not enough, ideal packing material must be sturdy, lightweight, and resistant to heat and corrosion.

Small-scale still builders may be tempted to use available alternative materials such as bronze scrubbing pads, stainless lathe turnings, marbles, or small sections of non-ferrous pipe as packing to save money. This is acceptable for smaller-diameter columns of 3 or 4 inches diameter, but with columns larger than that, the densely packed materials begin to create pressure drop up the column, which requires substantial amounts of heat energy below to overcome. While many of these materials offer considerable surface area, they're too tightly woven or clustered to provide the free gas space needed to encourage vapor flow. Unless the vapors can condense and revaporize freely — a phenomenon that occurs many thousands of times on their journey through the column — the enrichment process can't function to its fullest potential and alcohol yield will be reduced.

There is a definite pecking order with these alternative packings. Broad lathe chips perform better than scrubbing pads, are less costly, and tend not to clog up as much with free mash particles in the lower part of the column. Marbles are inexpensive, somewhat heavy, and offer only about 38 percent free gas space. The least expensive of the pall rings cost around $50 per cubic foot. Bronze scrubbing pads cost nearly as much, while lathe turnings can be had for next to nothing. From a long-term cost standpoint, in larger diameter columns particularly, it will pay to invest in commercial packing for the better performance it offers and especially for the energy saved throughout each run.

Condensers and Heat Exchangers

Technically, condensers and heat exchangers in the column share similar characteristics in that they both present a cool surface designed to come into contact with a hot medium. In the case of the condenser, its purpose is to reduce ethanol from its vapor phase to a liquid phase at the very top of the column — in other words, to condense the distillate vapors so they can be collected in a jacket surrounding the cool element and piped to a storage tank. Heat exchangers also cool vapors within the column, but they're used as means of heat control to keep temperatures at a specific level in one location. Still other types of heat exchangers or jackets add heat to a boiler or cooking vat to raise temperatures.

Fig. 8.18: *The condenser at the top of a 6-inch column. The hoses are part of the cooling water circuit.*

The elementary condenser of moonshine lore was just a coil of copper tubing set into a cold water bath. The alcohol distillate condensed within the tubing and was collected from the end. This arrangement was problematic in that mineral and protein deposits

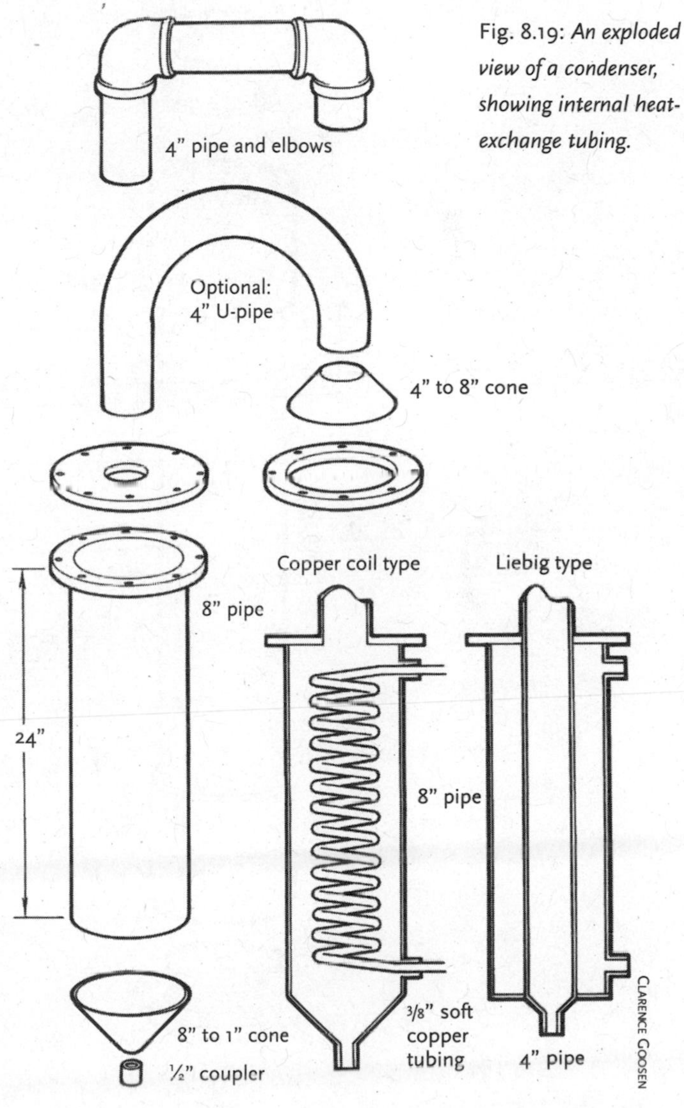

Fig. 8.19: *An exploded view of a condenser, showing internal heat-exchange tubing.*

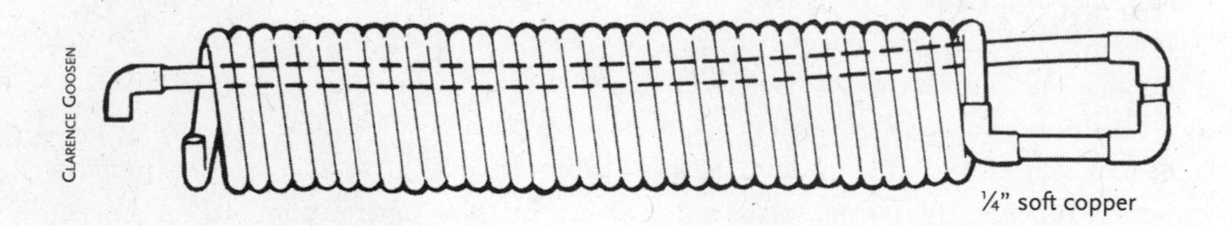

Fig. 8.20: *The condenser coil for a small-diameter column.*

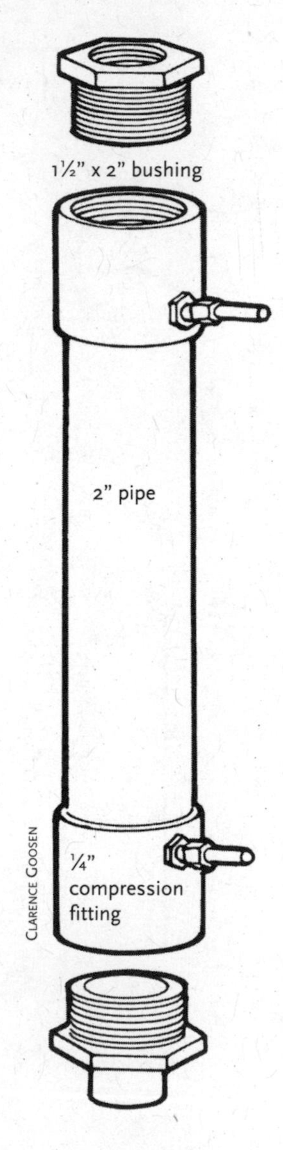

Fig. 8.21: *The condenser assembly, showing the coil inlet and outlet penetrating the jacket through fittings.*

would eventually clog the line, causing hazardous backpressure in the column and boiler, often causing an explosion with consequent injuries.

Modern condensers are much safer in that the cold water is run through the tubing and the condensate forms outside it in a larger foul-proof housing. Several feet of ¼-inch soft copper tubing is coiled around a section of pipe, which is used as a form, then the pipe is removed from the coil. The coil is then plumbed through the walls of the housing and sealed at the entrance and exit points. An even more efficient design is the spiral counterflow type, in which the alcohol vapors are sent through a copper tube with a diameter large enough to discourage clogging. The tube is wrapped with copper wire or strip in a spiral pattern, which promotes a swirling action in the cool water passing through a jacket surrounding the tube. This assures more contact and thus greater heat exchange than a conventional condenser.

A variation on the coil condenser is the Liebig condenser, in which the pipe carrying alcohol vapors is wrapped in a larger pipe jacket that contains the cold water. Though simpler, it is not quite as effective as the coil type because it has less contact surface area and, if made of thinwall tubing or pipe, does not offer the excellent heat-transfer attributes of copper.

Cold-coil heat exchangers within the column are also made of soft copper tubing formed around some type of cylinder. The coils in the column are larger in that more linear footage of tubing is used, along with a larger-diameter 3/8-inch tube. Once tightly coiled, the wraps of tubing should be drawn out and expanded, then re-arranged without kinking to fill the entire cross-section of column and give maximum exposure to the internal vapors. The heat exchangers are anchored to the column through ½-inch couplings welded through the column walls. Brass ½-inch pipe adapters and 3/8-inch compression fittings secure the tubing to the couplings.

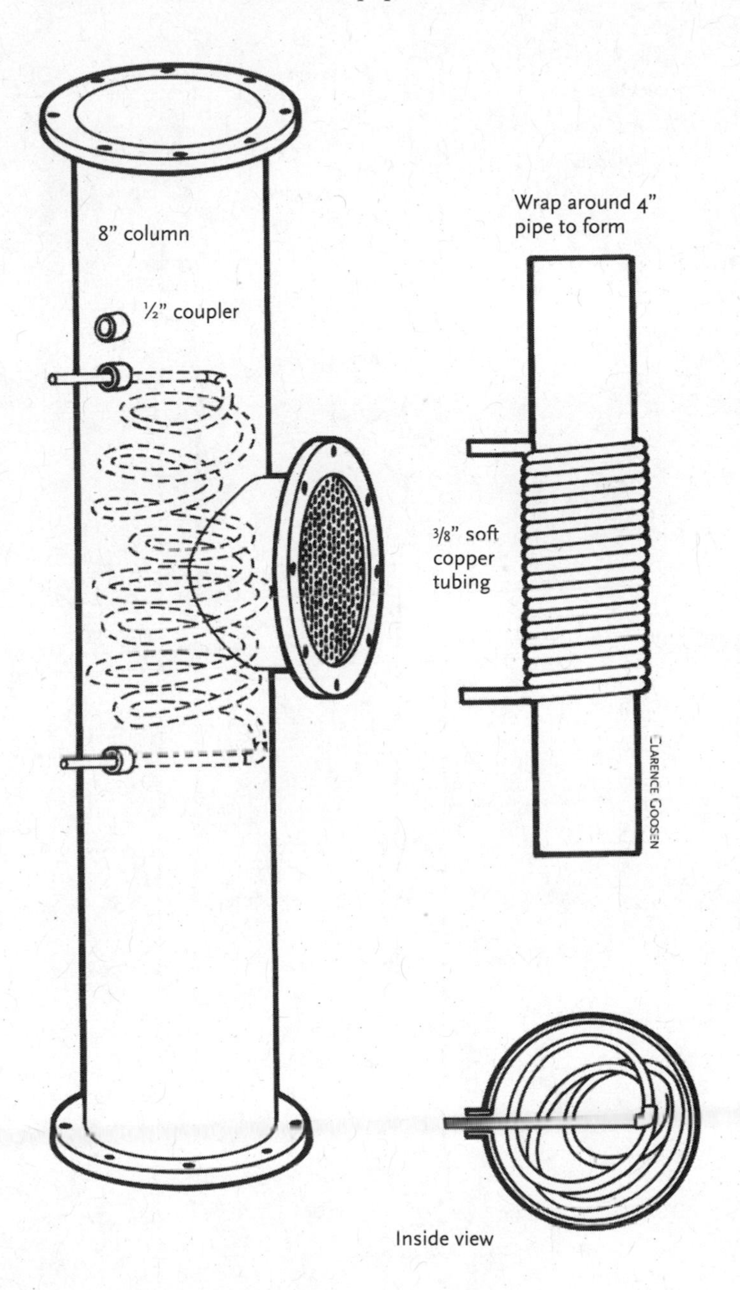

Fig. 8.22: *An exploded view of a typical mid-column heat exchanger.*

Dephlegmator

A dephlegmator is a device that cools the vapor mixture at the top of the column, condensing it and allowing liquid reflux to fall back through the column to the boiler tank, where it is revaporized and returned through the packing at higher proof. It functions much the same as a heat exchanger within the column, except that it's a separate unit rather than internal. Dephlegmators used in industrial applications are quite sophisticated and employ stringent temperature controls, but small-scale still designers can adapt the technology with reasonable success.

The modified dephlegmator is particularly suited to a small stripper column because, if controlled properly, it can rectify 100-proof ethanol to a 170-proof fuel without excessive energy input. You can think of it as an add-on

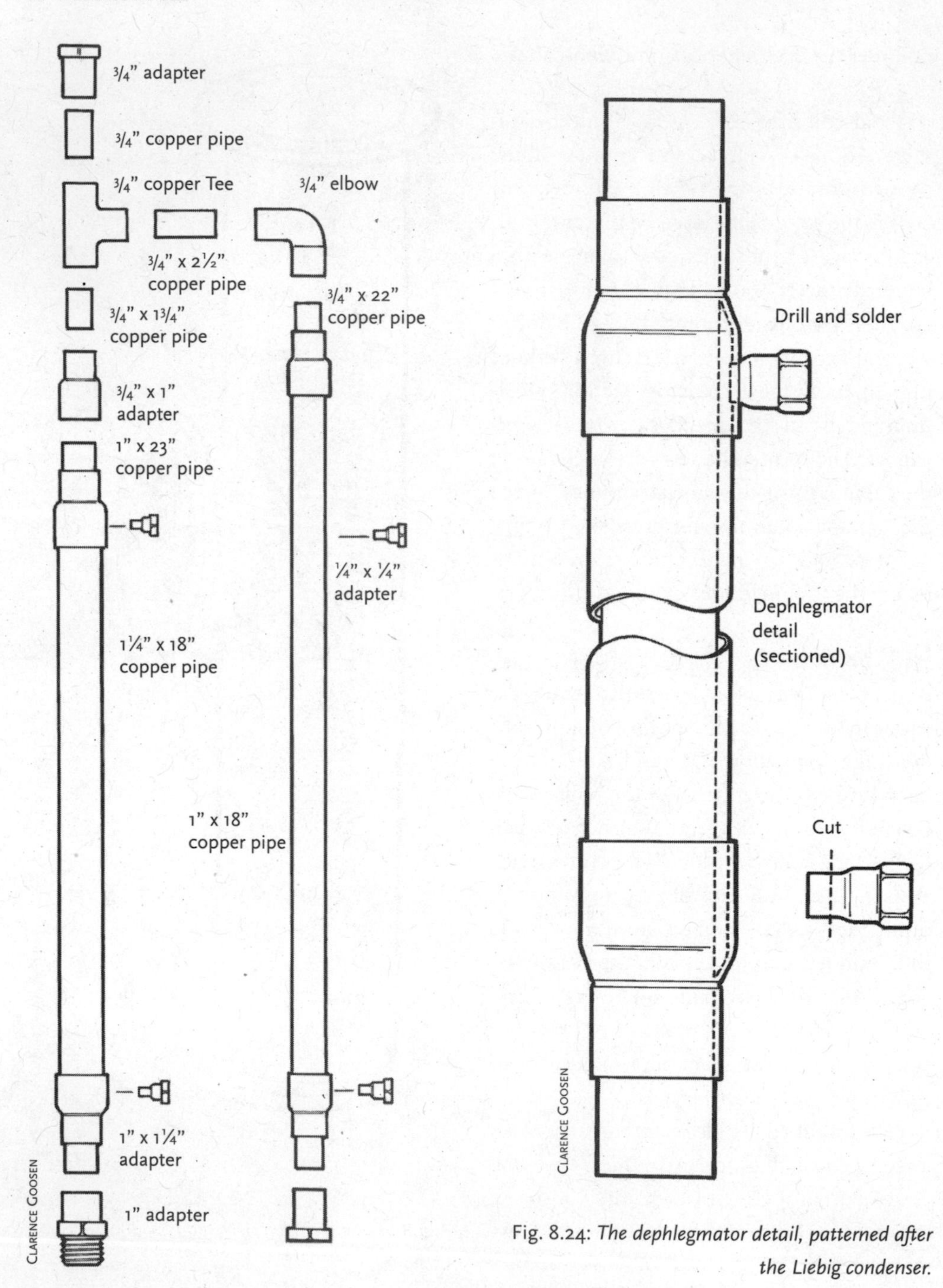

Fig. 8.23: *The dephlegmator condenser used in a 3-inch column design.*

Fig. 8.24: *The dephlegmator detail, patterned after the Liebig condenser.*

device that replaces the last few plates in the column. This particular design, developed by ethanol researcher Clarence Goosen for a 3-inch vacuum still he developed, uses a simple dephlegmator in concert with a condenser to economize on cooling water.

The dephlegmator is mounted at the very top of the tower and is plumbed directly to a condenser mounted adjacent to it. Cooling water first flows through the condenser jacket, then moves on to the dephlegmator jacket to be recycled. The rate of water flow determines the upper column temperature, which is strictly monitored with a temperature sensor mounted in a thermometer well built into the top of the dephlegmator. By maintaining the top of the column at its lowest temperature without actually condensing the alcohol vapors, the device keeps the upper portion of the column wet, ensuring fractionation and allowing the high-proof vapors to pass on to the condenser.

Figure 8.24 shows how the dephlegmator is assembled from copper pipe sections and fittings, following the design of the Liebig condenser. The core of the dephlegmator is too narrow in the small-diameter column to fit pall ring packing effectively, so stainless steel lathe turnings or non-ferrous scrubbing pads are used as packing.

The Reboiler

A reboiler is a chamber at the bottom of the stripper column that heats the column but allows boiler vapor to be introduced at the midpoint of the column instead of at the bottom. The advantage to this arrangement is that the entire column gets preheated and stabilized before any alcohol is introduced to it, but more significantly, it allows the initial alcohol-rich mash vapor mixture to ascend the rectifier or upper portion of the column without having to work its way through the lower part of the column. It's also a simple way of avoiding having to separate solids from the liquid mash to prevent clogging the packing in the column. When mash is fed into the column as a liquid, it usually has to be boosted to a minimum of 150°F before being introduced to the column. The preheated vapors eliminate this step and help the column come into equilibrium much more quickly than it normally would.

Reboilers are normally used for continuous distillation, where the rate of mash feed and the volume of liquid in the reboiler are coordinated, based on the rate of alcohol production in the column. In stills with separate stripper and rectifier columns, the volume of rectifier condensate is controlled in the system as well. But some batch stills use reboilers too, for better heat control. Elementary designs heat the reboiler by submerging it in the hot mash within the boiler tank (remember that the reboiler is sealed at the bottom, so no mash can enter unless it comes from the column above). More elaborate designs expose the bottom of the reboiler to direct heat from the firebox in a wood-fired system, or they use a designated controllable heat source such as a gas burner or an electric element. Using mash itself as a heat source is somewhat problematic in that its alcohol content is constantly changing during distillation, altering in turn the temperature at which the mixture boils due to the differential boiling points of ethanol and water.

Refluxing

Refluxing is the process of returning a portion of the condensed alcohol to the top part of the rectifying column. It's really a means to control temperature at that point, but it also has the benefit of keeping the upper packing or top plates moist with liquid to encourage fractionation.

Technically, refluxing refers to direct refluxing, e.g., physically spraying or pumping liquid ethanol into the top of the column. The cooled alcohol is dispersed through a perforated annular ring, where it mists and dribbles down through the packing. But for our purposes, we'll include the condensation of alcohol vapor by an internal heat exchanger as refluxing, as well. The heat exchanger is located at the top of the column, and though it does consume a significant amount of water (which you should cool and reuse, or recycle hot in some way), it saves having to buy a separate alcohol reflux pump and controls.

Fig. 8.25: *Condenser water controls can be inexpensive needle valves. Whatever type of valve is used must allow fine control.*

The vapor temperature at the top of the column will be your indication that things are going well ... or not. If it maintains a steady state and vapor is indeed passing through to the condenser, all is well. Once the temperature starts to rise, you'll have to cool the exchanger coils (or add more alcohol in a direct reflux system) until it stabilizes again. If the temperature drops, you'll need to cut back on the cooling water (or alcohol) accordingly.

Practical Layout

It's difficult to outline the "best" layout plan for your distillery, but using the progressive flow of how alcohol is made as a guideline seems to work well. Figure 8.28 shows a typical flow pattern, which I'll describe here. Feedstock processing and storage can occur outside the actual distillery area or building, under a roof or in an outbuilding if possible. Within the distillery building itself, the cooker can be set up at one end, with fermenting tanks, if used, alongside one wall. If the prep work is done in the same building, that area can be set up at the same end as the cooker.

The distillation tank or column should be in the vicinity of the cooker and fermenting tanks with the plumbing directed toward the handling equipment such as the mash pump and transfer pumps. Hot water storage should be placed as near to where it will be used as possible; cooling water and fuel alcohol collection tanks can be set further away.

You can expect spills and leakage of mash and slurry, so be prepared to handle cleanup

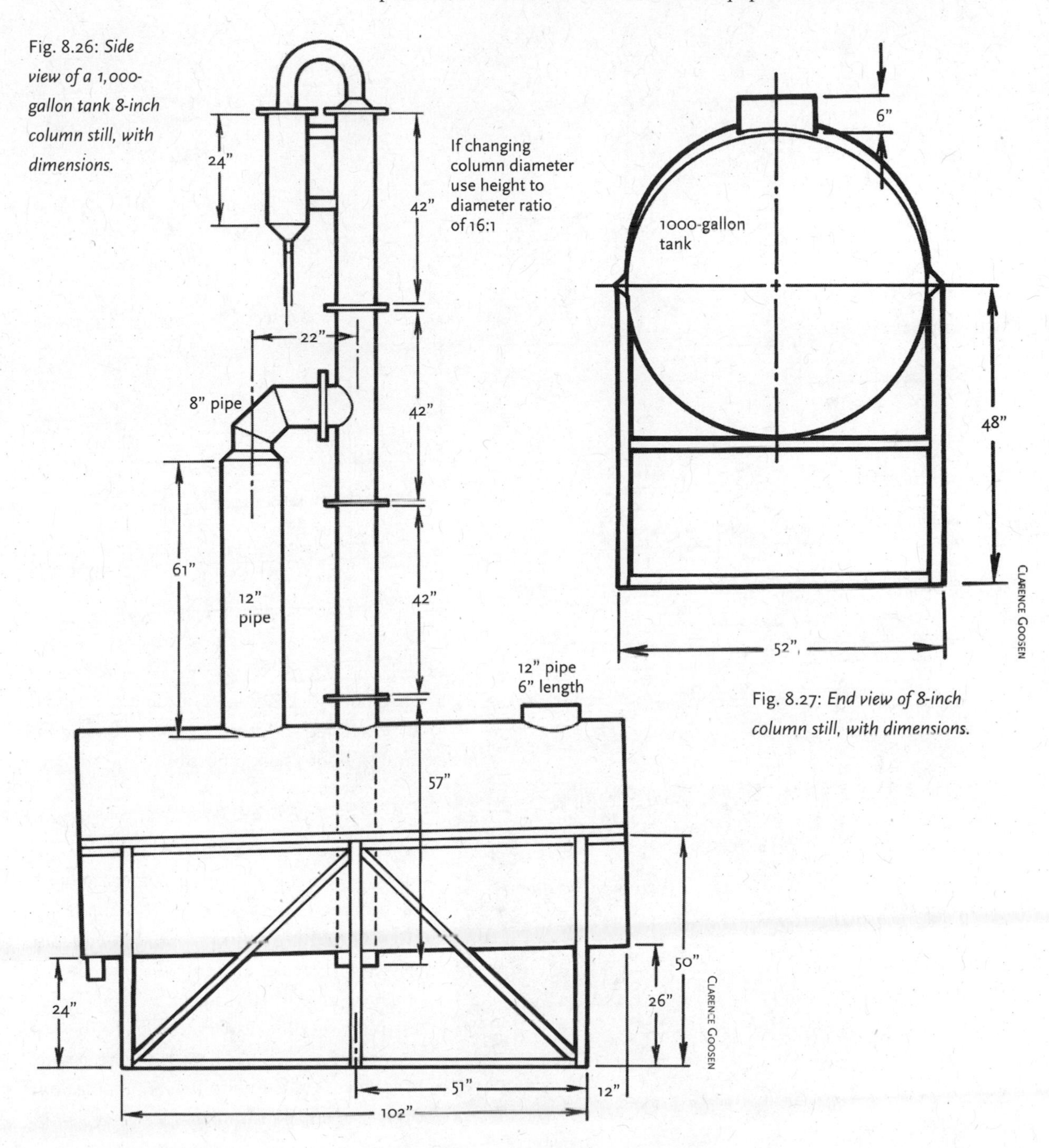

Fig. 8.26: *Side view of a 1,000-gallon tank 8-inch column still, with dimensions.*

Fig. 8.27: *End view of 8-inch column still, with dimensions.*

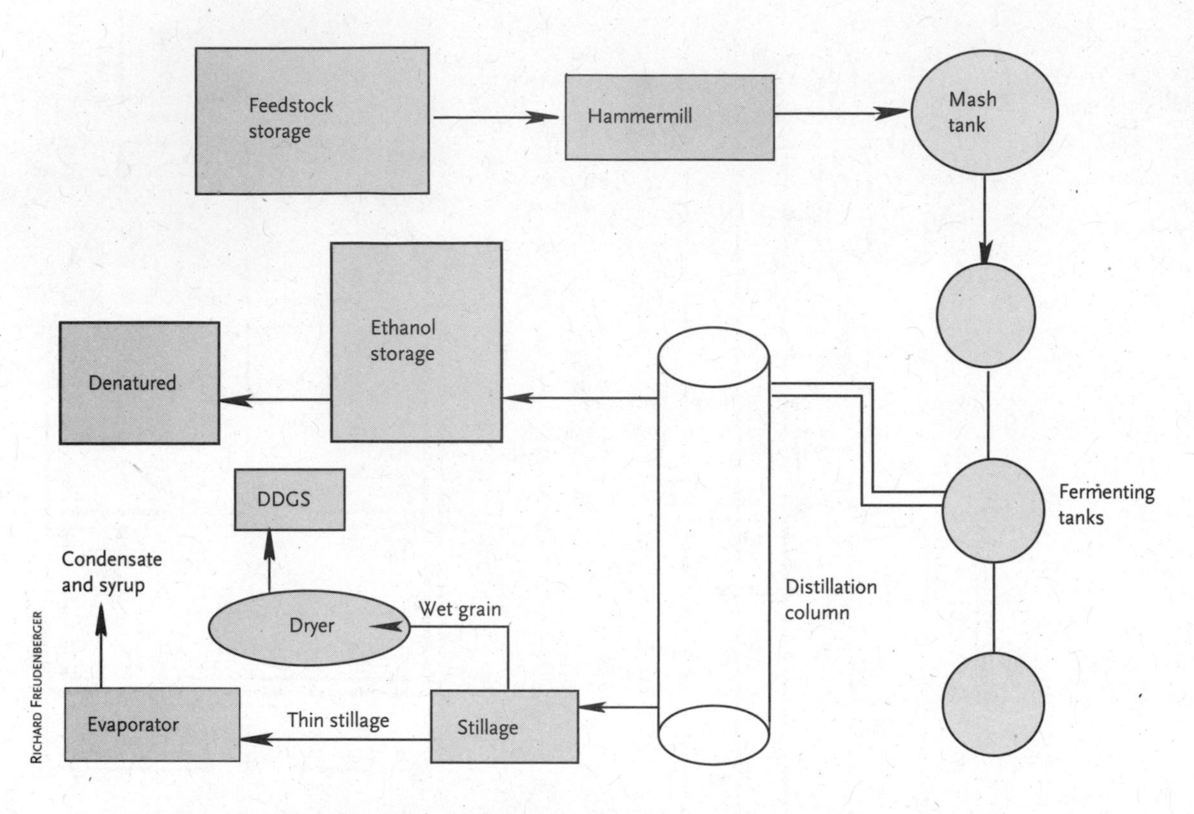

Fig. 8.28: *It's best to develop a working flow pattern before actually setting up your distillery, regardless of how small it may be.*

Fig. 8.29: *Preparing to test proof strength of fuel alcohol from a small batch still.*

of liquids by leaving access around and behind all processing and storage tanks, and incorporating a floor drain, if possible.

Operating a Batch Still

A batch-run still with a packed column is the easiest type of distillation equipment to operate because there is very little to go wrong, given a healthy supply of mash. If you do miscalculate or make an error in heat control, you'll be able to start again, without losing anything but time. Experience will be the best teacher, of course, but here are some general procedural guidelines to help in getting your operation up and running.

Preparing the Tank

With some stills, depending upon their heating method and column design, the mash can be run complete with solids. However, the simpler packed-column stills do not handle solids well because the particulates either clog the packing, or more likely, get scorched at the bottom of the tank in heating. Unless you have an internal agitator or some other method of keeping the solid particles in suspension while the liquid's being heated, you'll need to separate the solid distiller's grains from the mash to create a liquid beer that's in the range of 10 percent alcohol but free from particulate material. Dewatering devices were addressed earlier in this chapter, but for your early "getting used to it" runs, straining, pressing, or other simple and inexpensive means of separating the liquid from the solids is adequate.

Before filling the tank, all valves and cleanouts should be closed and sealed. If the boiler tank is separate from your cooking vat, be certain that the tank is clean and free of residue.

Initiating and Controlling Heat

Start up your heat source. With a gas or electric burner, you'll have more finite control over temperature, but even with wood you'll be able to — and, in fact, will have to — control temperatures in order to run the still for capacity or higher yield. What's the difference? It is simply that in the lower heat ranges the liquid beer will merely simmer, keeping the vapor temperature low and allowing vapors to rise slowly and predictably through the column. With higher heat, the beer will boil vigorously and the vapors will rise into the column accordingly, so you'll have to maintain a close watch on the column's heat exchangers to keep it in equilibrium, or your proof level will suffer.

Not surprisingly, the higher boiling rate requires sending a greater volume of water through the heat exchangers to draw away excess heat, though it does increase the still's capacity. Low heating conserves water and keeps proof levels up, but distillation takes longer. If heat input is *too* low, however, the alcohol will not vaporize in the column and therefore can't be condensed to liquid, so production will halt until temperatures are returned to a working level. At the other end of the scale, from both an economic and environmental standpoint, you should make every attempt to recycle or put to use the warm discharge water from the heat exchangers. It's very suitable for preheating the next batch of mash or a domestic hot water supply, or for watering livestock in an agricultural environment.

Maintaining Column Equilibrium

Temperature at the midpoint of the column should be maintained at approximately 185°F, the boiling point of a 50 percent equal mixture of ethanol and water. Depending on the design of the column and the packing, the exact ratio at midpoint may vary, so the temperature may have to be varied slightly as well. The temperature at the top of the column needs to be maintained at a level just above the vaporization point of nearly pure alcohol, which is 173°F. For reference, Table 7.1 in the previous chapter shows the vaporization temperatures for various alcohol-water blends.

With a batch still, however, the concentration of alcohol in the beer decreases as ethanol is drawn off further into the run, so the boiling point of the beer will rise and will continue to do so until the mixture is nearly all water and boils at 212°F. To keep the column in equilibrium, you should find a temperature that works with your design and try to maintain it throughout the run. Do not alter flow through the heat exchangers abruptly, as this will swing temperatures too far beyond your target. It's better to make minute adjustments and allow time for them to take effect before making a second or larger adjustment. It only requires a slight opening of a water valve to increase the flow of cold water, but the water must travel through many feet of tubing, conducting heat from the column the entire way, before it can establish a new equilibrium. The thermometer or temperature gauge at the column will indicate the results of your changes and it should be monitored carefully.

Too much heat (or not enough cooling) in the column will yield a low-proof product. Too much cooling in the column lowers temperatures below the vaporization point of the alcohol, and the vapors will condense at the heat exchanger coil and fall back down the stripper section and into the boiler tank (or reboiler, if equipped), never getting an opportunity to rise past the cool coil. More sophisticated designs use automatic flow valves with thermocouple sensors to control cooling flow, but any temperature changes still have to be made incrementally.

Keeping a watchful eye on the proof of the condensate collected from the condenser will help you to establish where temperatures need to be to make a run consistent and productive, and with each run you'll gain more experience.

Finishing Off and Low Wines

The last quarter of the run will show some marked differences from the first three-quarters. As I mentioned earlier, the boiling point of the mixture in the tank will rise as more alcohol is evaporated, condensed and drawn off in collection. Near the end of the run, temperatures in the boiler will be close to 210°F, and maintaining a temperature of even 185°F at the top of the column becomes more difficult. A lot of cooling water will be consumed to remove water from the mixture, and it will take as long to draw off a high-proof product from the final quarter of the run as it did to get the same or better product from the first three-quarters.

From an energy and time perspective, the best course of action at this point is to allow the column temperature to rise and do your best to maintain the midpoint column temperature several degrees below that of the boiler temperature, and the top column temperature several degrees below what the midpoint is. This will allow a portion of the mixture to reflux, or separate into water and ethanol, and will expedite the process significantly, saving valuable heat energy. The proof levels in this final stage will be relatively low, but there is enough alcohol in the product to make it worth saving. These so-called "low wines" can be stored in a separate vessel and combined with the low wines from successive runs to make a whole new distillation run in the future. As with the main distillation run, you will also create a substantial amount of hot water, which should be reused in a productive manner.

Cleanup and Maintenance

It's never a good practice to let your tanks and equipment sit uncleaned after a run unless you're planning a second run immediately thereafter. Bacteria will proliferate in a warm and moist environment, making it all the more difficult to effectively clean vessels and plumbing when you need to. Secondly, moisture and even slightly acetic conditions will be corrosive to mild steel over time (this isn't as much an issue with stainless steel). The best practice is to remove any spent residue from the tank(s) and pressure-wash the insides with hot water. Pumps, plumbing and fittings can be washed with chlorinated pipeline detergent and rinsed thoroughly so no chlorine residue remains to affect the next run. Dry all metal parts with compressed air if air-drying isn't feasible.

Vacuum Distillation

Under normal conditions, at sea level, there are about 14.7 pounds per square inch of air pressure bearing down on the earth's surface. At higher elevations, there is less air available to press down, so the pressure isn't as great. It makes a difference: As anyone who's lived at mile-high elevations can attest to, liquid boils at a lower temperature than it does in the flatlands, and that can create problems in cooking certain foods.

Distillation in a vacuum, in which air pressure is maintained well below normal

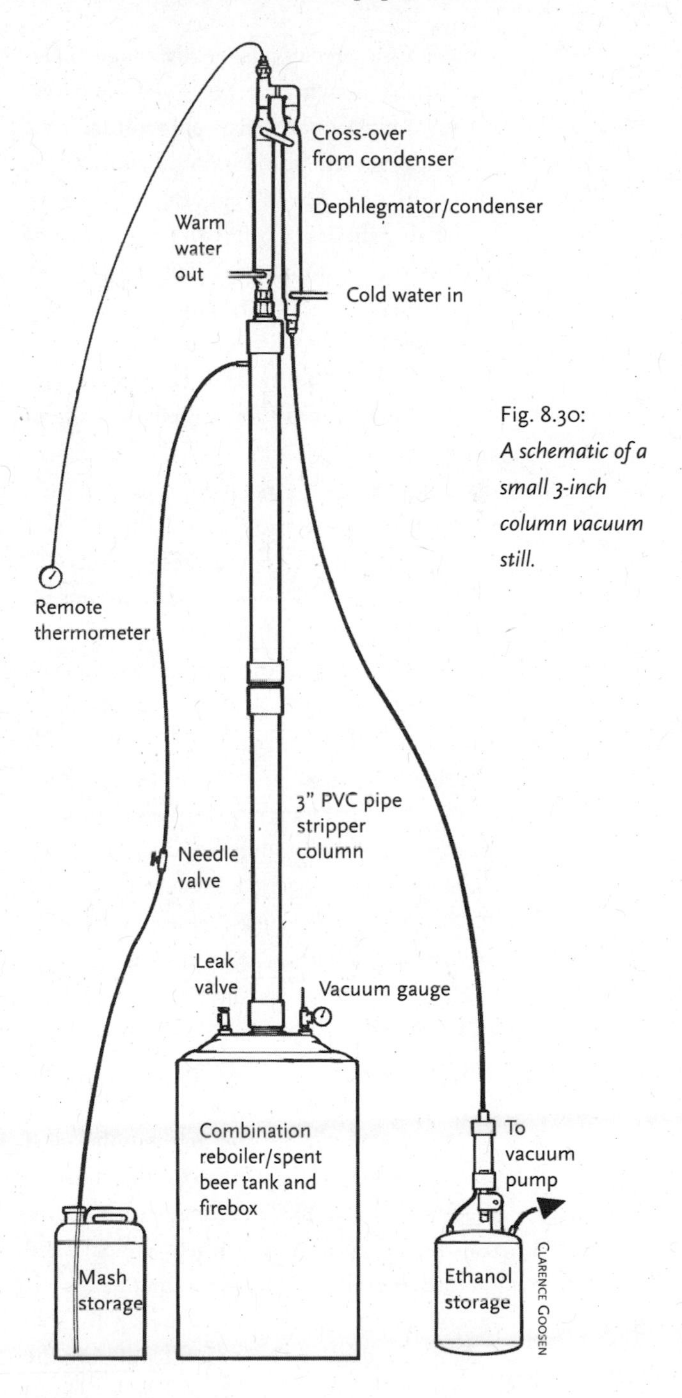

Fig. 8.30: *A schematic of a small 3-inch column vacuum still.*

atmospheric pressure, takes advantage of this fact. When distilling alcohol, we're not necessarily seeking a particular temperature. Our goal is to get the liquid boiling with as little energy input as possible (in heat or Btu's). So, vacuum distillation offers the advantage of energy conservation, plus — once the system is brought into equilibrium — it's easier to maintain in a steady state.

Let's look at this in detail, because distillation in a vacuum isn't as simple as it might appear, nor is it the solution to all ethanol production problems. First, it requires a vacuum pump of some sort. Commercial models used by refrigeration and cooling professionals cost hundreds of dollars — and even thousands for stationary units; compressors salvaged from box or chest freezers cost far less, but their oil seals can be damaged by alcohol vapors in the air. The best choice is an oil-less pump capable of drawing at least 0.5 cubic feet per minute and 25 inches of mercury (In. Hg), a measure based on its ability to pull the heavy liquid element up a tube (a perfect vacuum is just over 29.9 In. Hg by gauge measurement). That kind of vacuum pressure can be fatal to thinwall steel, so it's important that the column, tank and plumbing be substantial enough to withstand the stress. At 25 In. Hg, water boils at around 130°F and alcohol at about 106°F.

Water flow to the cooling chamber and condenser at the top of the column must be meticulously controlled to regulate temperature at that point, so a temperature-actuated modulating valve, rather than a manually controlled valve, is recommended. So is an accurate means of measuring temperatures at both the mash boiler and the top of the column. A reliable vacuum gauge for the tank is necessary, as well as a fine-control needle valve to introduce atmospheric air into the system for initial heat-up and to control vacuum if needed. Finally, from the standpoint of safety, a pressure safety release valve should be installed on the boiler tank of this (and any) distillation vessel to relieve pressure buildup if the vacuum pump should fail. A 15-psi limit is safe; standard water-heater temperature and pressure valves may have too high a pressure threshold (150 psi) and a marginal temperature limit (210°F), especially for atmospheric distillation.

Low-temperature distillation opens the door to solar thermal preheated water and even direct solar thermal water if concentrating collectors are used. Other forms of renewable or biomass-based energy might be practical here, where they would not be otherwise — for example, an electric-element heat dump from a wind or microhydro turbine, or co-generated heat. Lower temperatures also pose less operating risk from accidental burns, which can be serious at normal boiling levels.

Vacuum distillation, on the other hand, comes with reduced capacity unless you opt for a continuous distillation process. You can expect up to 40 percent less capacity per hour for a given size batch column under vacuum, meaning that initial material costs will be greater from the outset to achieve equivalent yields. Continuous distillation means that you'll have to filter your mash before introducing it to the column, because the packing easily becomes clogged with solids. Controlled

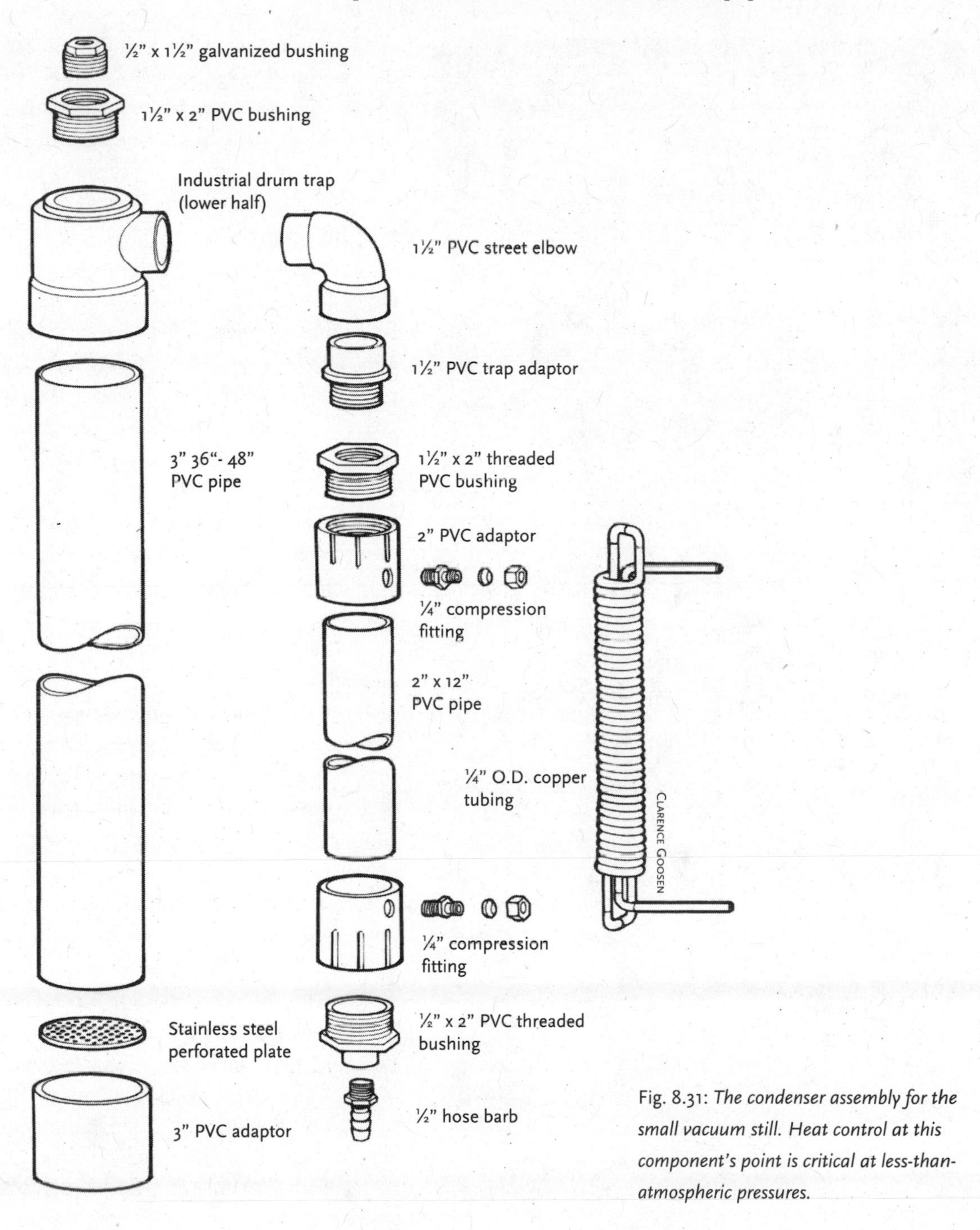

Fig. 8.31: *The condenser assembly for the small vacuum still. Heat control at this component's point is critical at less-than-atmospheric pressures.*

CLARENCE GOOSEN

Fig. 8.32: *A working demonstration model for vacuum distillation. Because of low operating temperatures, the column is fabricated from plastic pipe.*

cooling under vacuum also requires a substantial amount of water because the temperature of the vapors at the top of the column isn't that high to begin with, so the temperature differential is narrow. Either a lot of water flow or very cool water input is needed to draw down the temperature to working levels.

Operating a vacuum still involves, by and large, getting the column into equilibrium. Firing the reboiler with water in a normal atmosphere allows it to build up heat to the point where it will boil at reduced atmospheric pressure. When that happens (at a temperature around 130°F), the vacuum pump can be activated and the atmosphere drawn down to or near the desired level of 25 In. Hg, at which point mash can be introduced to the column at midpoint. From here, heat control through the reboiler and via the top coil is critical. Too much heat in the reboiler overheats the column and delivers a low-proof distillate. Too much cooling at the top causes premature condensation and prevents alcohol vapors from passing into the condenser; they simply reflux and fall back down the column. Conversely, not enough cooling at the top creates the equivalent of an overheated column. The vapor mixture will not fractionate and low proof will result. The mash feed rate must also be coordinated to optimize with the rest of the system or the column will flood and fractionation will not occur.

The question of whether vacuum distillation is worth the investment expense goes back to how "appropriate" you want your distillery to be. If your fuel costs are high and you're running a continuous process on a regular or frequent schedule, then an artificial atmosphere still may have some cost-saving value to you, even given the additional initial expense. If you're a casual fuel producer operating a batch still on the scheduled occasions when you need fuel, vacuum technology makes considerably less sense.

Alcohol Storage

Bulk storage after the fuel has been denatured can be above or below ground. If you feel that a standing tank might be a target for theft, it can be protected with a simple enclosure such as a panel or chain-link fence. Underground storage has the benefit of being out of sight and out of the way. Unfortunately, the incidence of underground leakage, especially from service station tanks containing gasoline with the carcinogenic MTBE additive, has prompted state and federal environmental agencies to regulate buried fuel storage well beyond any practical return, albeit for the protection of our groundwater. Single-wall steel tanks are particularly prone to the corrosive effects of wet soil, and leaks are very difficult to detect deep in the soil.

That leaves above-ground storage, which is actually less of a risk to the environment because any leak or damage is obvious and can be repaired promptly. The fact that 190-proof alcohol has a flashpoint — the lowest temperature at which vapors can ignite in air with a source of ignition — over 100 degrees higher than that of gasoline, (which is still dangerous at minus 45°F) indicates that it is safely stored above ground.

The simplest type of tank is the horizontal cylinder mounted on a steel ladder frame several feet from the ground. The fuel from these tanks is gravity fed through an automatic shut-off nozzle, usually secured through a ball valve on the tank outlet. The frame's four feet should be bolted to substantial concrete pads set in the ground, preferably 16 inches square and excavated to the frost line for your region. Five-hundred gallons of fuel ethanol weighs over 3,300 pounds, enough to sink an unpadded leg deep enough into wet ground to cause the frame to buckle. It would be a good idea to check with your local code jurisdiction to see if any other requirements have to be met, such as earthquake protection or spill containment. A full-size slab with a short perimeter knee-wall may be required.

Other than a fuel delivery fitting, the tank should have a 2-inch fitting for the fill cap and drop tube, and another one for a combination pressure-vacuum vent, sometimes called a conservation vent. This is a dual-action device that safely releases excess pressure in the tank in hot weather, and prevents outside air from being sucked into the tank when it cools until a significant vacuum is formed. A mechanical fill-level gauge is also a convenient item. If possible, avoid setting your tank in direct sunlight, or at least provide some type of shade screen or lean-to that is non-combustible.

If a frame-mounted tank isn't feasible for aesthetic or safety reasons, you'll have to use a pad-mounted surface tank. The pad should be reinforced concrete and the tank supported off the slab on the short metal feet attached to it. Because you won't have the benefit of gravity here, one of the 2-inch fittings will have to accommodate either a crank-operated or electric suction pump with a tube and pickup strainer. Besides any grounding associated with the breaker box for an electric pump, the tank itself must be grounded, as well as the filling nozzle, which should be grounded to the tank, unless it's non-metallic. Static electricity can build up in both steel and reinforced plastic tanks and must be sent to earth to prevent any risk of sparks during handling.

Fuel alcohol, even manufactured at a small plant, is not supposed to be stored in containers of less than 5 gallons in size except for labeled testing samples, according to federal TTB regulations. Also, any alcohol fuel container of less than 55 gallons should be designated with a label that states "Warning — Fuel Alcohol — May Be Harmful or Fatal if Swallowed." For safe drum storage, do not store lower-proof alcohol fuel in steel drums, as they can corrode at the seams from the inside. Higher-proof ethanol can be stored in steel drums, but only if they're positioned vertically, with the fittings at the top, as the threads tend to leak.

It's always a good idea to run the fuel through a filter before pumping it into the vehicle tank. Three-micron and even one-micron filters are the best choice, as they'll remove even the finest contaminate particles, but be sure you choose an alcohol-tolerant filter; most conventional fuel filters are designed to separate and absorb water and will eventually become plugged, especially with alcohol below 190 proof.

NINE

Alcohol as an Engine Fuel

Combustion properties • Emissions factors • Performance and fuel economy • Engine modifications • Factory Flexible Fuel Vehicles • Fiat's TetraFuel system • Alcohol and diesel engines • Space heating systems

For decades, the cheap and easy availability of gasoline has led us to believe that the petroleum-derived fuel is the only one suitable for internal combustion engines. The petroleum industry has dedicated itself to promoting its products, and, using a combination of skillful marketing and successful government lobbying, it has built a massive transportation and industrial infrastructure on derivatives of oil.

Yet, as explained in Chapter 1, gasoline is only one of several fuels that have proven themselves to work in a gasoline engine, and in fact kerosene, propane, natural gas and ethanol have all at one time or another served as dependable fuel sources. Aside from the limitations of infrastructure (which is a significant reason why alternative fuels can't seem to get traction in the marketplace), a large part of the reluctance to consider fuel alternatives is simply that auto engines have been optimized for gasoline. Changing that is not impossible, and not even difficult, but clearly would require some investment and incentive, which doesn't always come easily.

Combustion Properties

Gasoline is a complex mixture of hydrocarbons — molecules made up of hydrogen and carbon — and other elements such as sulfur, nitrogen, boron and phosphorus. It is a combination of dozens of different compounds, which vary according to the source of the crude oil from which it's made. When the crude oil is heated and distilled at the refinery, it is separated into different products according to its intended use: petroleum gases and solvents, aviation gasoline, automotive gasoline, kerosene, heating oil, diesel fuel, lubricants and waxes, and furnace oil and asphalt. The compounds that boil at lower temperatures, such as gasoline, are lighter, and those with the higher boiling points — lubricants and furnace oil — are heavier.

CLARENCE GOOSEN

Fig. 9.1a: *A tidy alcohol conversion on an older VW beetle uses air preheating as part of the technique.*

CLARENCE GOOSEN

Fig 9.1b: *The author with the VW-powered Pober Pixie experimental aircraft, converted to alcohol fuel at an Experimental Aircraft Association Fly-In in the early 1980s.*

There is no "typical" gasoline formula because refineries further vary the blend to suit seasonal changes, altitude, emissions requirements, and so forth. But, for a representative sample, we can say that octane is as good an illustration as any, with its 8 carbon atoms and 18 hydrogen atoms (C_8H_{18}). Ethanol, as you've learned, is also distilled from carbohydrates, which contain carbon, hydrogen and oxygen atoms. With ethanol, one of the hydrogen atoms has been replaced with a hydroxyl radical, which is an oxygen atom bonded to a hydrogen atom. The molecular formula for ethanol — C_2H_5OH — reflects this. Ethanol is only one of many different types of alcohol, but only ethanol and methanol (a toxic alcohol derived from wood distillates or synthesized from natural gas) are normally used as fuel. Methanol is popular in sprint and drag racing, and it was the fuel of choice of the Indy Racing League from the mid-1960s until 2007, when it was replaced with 100 percent ethanol. But methanol is only mentioned here as a point of discussion — it is not economically feasible to make on a small scale, and it is toxic even in small quantities or when absorbed through the skin — hardly the ideal home-shop project.

The differences between gasoline and ethanol are abundant enough to comment on, and not simply because of the oxygen present in ethanol. Table 9.1 compares the properties of the two, and in the following few sections I'll point out a few points of particular interest.

Heat of Combustion

A fuel's potential energy is measured by the amount of heat released by a given volume or weight of fuel when it completely combusts with oxygen. It's sometimes referred to as energy density and is measured in British thermal units (Btu's), defined as the energy required to raise the temperature of one pound of water one degree Fahrenheit, the equivalent of about 252 calories. With fuels, the heat of combustion is generally presented as its lower heating value. The lower heating value of a typical gasoline sample is 111,000 Btu per gallon, while the heating value of ethanol is 76,500 Btu per gallon.

Clearly, gasoline has a Btu advantage of over 40 percent, due largely to the fact that ethanol has a significant oxygen content by weight. The presence of oxygen leans out the fuel mixture, which is why more fuel must be added to the air/fuel ratio in compensation when burning ethanol. At the same time, its presence also

Table 9.1
Liquid Fuel Characteristics of Gasoline and Ethanol

Chemical Properties *Formula*	**Gasoline** *complex mixture*	**Ethanol**
Molecular Weight	variable	46.07
Percent Carbon (by weight)	85-88	52.14
Percent Hydrogen (by weight)	12-15	13.12
Percent Oxygen (by weight)	variable	34.74
Carbon/Hydrogen Ratio	5.6 - 7.4 : 1	4.0 : 1
Stochiometric Ratio (air-fuel)	14.2 - 15.1 : 1	9.0 : 1
Physical Properties	**Gasoline**	**Ethanol**
Specific Gravity	.70 - .78	.7936
Liquid Density (Lb./Cu. ft.)	43.6	49.3
Liquid Density (Lb./Gallon)	5.8 - 6.5	6.59
Boiling Point (Degrees F.)	80 - 440	173.3
Freezing Point (Degrees F.)	minus 70	minus 174.6
Solubility (in water)	240 ppm	miscible
Solubility (water in)	88 ppm	miscible
Vapor Pressure at 1 Bar (100°F.)	7 - 15 In. Hg.	2.5 In. Hg.
Vapor Pressure at 1 Bar (77°F.)	.3 In. Hg.	.85 In. Hg.

Table 9.1 cont.
Liquid Fuel Characteristics of Gasoline and Ethanol

Thermal Properties *Formula*	**Gasoline** *complex mixture*	**Ethanol**
Heat of Combustion (77°F.)		
Lower Heating Value (Btu/lb.)	18,900	11,550
Lower Heating Value (Btu/gal.)	115,400	76,114
Higher Heating Value (Btu/lb.)	20,250	12,780
Higher Heating Value (Btu/gal.)	124,800	84,220
Latent Heat of Vaporization (77°F. at 1 Bar) (Btu/lb.) (Btu/gal.)	150 900	395 2,603
Flash Point (Degrees F.)	minus 50	55
Autoignition Temperature (Degrees F.)	430 - 500	793
Octane Rating (Research)	90 - 101	106
Optimum Air-Fuel ratio	15 : 1	9 : 1
Explosive Limits Air-Fuel Ratio	13.2 : 1 - 71.4 : 1	5.3 : 1 - 23.3 : 1
Explosive Limits in Air (by percentage)	1.4 - 7.6	3.3 - 19
Maximum Practical Compression Ratio (spark ignition)	9.2 : 1	15 : 1

makes ethanol an oxygenated hydrocarbon, a much cleaner-burning fuel than gasoline, which we'll address in detail further on in the chapter.

Heating value is far more important in, say, your home's furnace than it is in your car's engine because an internal combustion engine's effectiveness is measured in work done, not heat generated. In fact, the more heat generated by an engine, the less efficient it is because less energy is available to do the actual work of driving the pistons. Work, or energy, is measured internationally in *joules*, as the force required to move a kilogram (mass) one meter (distance) in one second (time). Power is the rate at which work is done, and it is measured in watts: the energy transferred by one joule in one second is one watt. In automotive terms, horsepower is a more common measure, and one horsepower is equal to 746 watts.

Within the engine's combustion chamber, how the fuel burns and the rate at which it burns is more significant than its heating value. Those characteristics are controlled not only by the fuel itself, but also by factors such as temperature, compression ratio, combustion chamber design, and how well the fuel is atomized, or broken into minute particles.

Peak combustion pressures are actually lower for ethanol than for gasoline, but the cylinder pressures remain higher for a longer period, so there is a greater crank duration available to the engine. The lower peak cylinder pressure also helps with detonation control.

Octane Rating

The octane number that you see on gas station pumps is a measure of the fuel's "anti-knock" qualities (detonation resistance), or its ability to withstand auto-ignition (also called "ping"). The fuel is measured against octane, the component of gasoline mentioned earlier, as a standard, which is given the value of 100.

In the compression cycle of a gasoline engine, a mixture of fuel and air is compressed by the moving piston, typically at a ratio of about 9:1 in contemporary engines. In other words, the volume inside the cylinder when the piston is at its lowest point is nine times greater than when the piston is at its highest point. The compression of the mixture in the cylinder by the piston heats the gases significantly — enough that temperatures can approach the point at which the fuel will auto-ignite, without any contribution from the spark plug. Auto-ignition is undesirable because it is not *controlled* ignition; the substantial pressures and localized heat that auto-ignition knocking creates puts considerable stress on engine components and can even burn holes in the piston.

Manufacturers set the compression ratio to suit the fuels available; most cars are designed to use regular unleaded and have moderate compression ratios. Higher-performance models are built to use high-octane gasoline so they can take advantage of higher compression ratios — in the range of 10:1 or 10.5:1 — for increased power. So, fuels with greater resistance to knock or auto-ignition offer greater flexibility in engine design.

The auto-ignition point of regular gasoline can range from about 500°F to as low as 430°F. Ethanol has an auto-ignition point of 685°F. The octane rating of the highest-grade automotive gasoline is about 93. Ethanol has an octane rating of 106. This allows it to tolerate exceptionally high compressions ratios — higher than most unmodified automobile engines can handle without destroying themselves over time. Ethanol can in fact abide compression ratios of 16:1 or so — but realistically, the optimal balance between high operating efficiencies and the cost of installing high-strength forged engine components is achieved at about a 12.5:1 ratio.[1]

High compression delivers increased power, greater mileage, and improved efficiency. But even without increasing the compression ratio, drivers can benefit from ethanol's high octane rating by simply advancing the engine's ignition timing. Because ethanol has a broad range of tolerable air-fuel mixtures compared to gasoline and burns with a rapid, broad flame front owing to its oxygenated structure, the effects of early detonation are not detrimental to the combustion chamber in the way that gasoline detonation is. Peak temperatures are reduced, and pressure is not concentrated in a single area.

Volatility

Volatility is the disposition of a liquid fuel to evaporate at a relatively low temperature. If vaporization doesn't occur within a temperature range available to the engine and its fuel-handling components, the carburetion or injection system cannot deliver the proper air-fuel mixture and the fuel won't combust efficiently or completely in the engine. Not only does this result in wasted fuel, but it significantly increases levels of carbon monoxide and hydrocarbon emissions.

A fuel's flash point is related to its volatility in that an evaporative compound such as gasoline is hazardous in the vicinity of sparks or an open flame. Ethanol, too, is highly evaporative, but its flash point — the temperature at which a combustible substance will ignite when its vapors are exposed to flame — is well over 100 degrees greater than gasoline's, reducing the possibility of unintended ignition. However, there is the factor of flammability limits that also comes into play. This has to do with the fuel percentage by volume in air that will support combustion. Gasoline has a fairly narrow range, with a maximum proportion of about 7 percent. Theoretically, this means that gasoline won't burn by itself if the mixture of fuel to air is any greater because there isn't enough oxygen to support combustion — the mixture is too rich. Ethanol has a larger point spread at a higher percentage — somewhere between 4 and 19 percent. So, in theory at least, it has more potential for ignition because of its broader flammability limits.

Latent Heat of Vaporization

When a substance is about to undergo a change in state, as in transition from a liquid to a vapor, it must absorb additional heat in order for the change to occur. This is known as the heat of vaporization. Ethanol requires 395 Btu per pound (at 77°F at 1 atmosphere) to make the change; gasoline requires about 147 Btu per pound. This difference, a factor of 2.68, is the basis for ethanol's cooler operating condition in an internal combustion engine.

Since the heat is removed from the closest source available — the cylinder head and intake manifold — what occurs is that the air-fuel mixture cools upon entering the cylinders. Gasoline, with its lower heat of vaporization, cannot absorb nearly the amount of heat that ethanol can, so the heat is simply transferred to the engine coolant as waste.

Furthermore, exhaust temperatures in an ethanol-fueled engine typically run cooler than in its gasoline-fueled counterpart, throughout the working rpm range. Tests at Baylor University Institute for Aviation Science on a 125-horse-power Lycoming O-235 piston-driven aircraft engine indicated an exhaust gas temperature (EGT) of 1,164°F on E-95 ethanol compared to 100-octane Low Lead gasoline's 1,326°F at 2,100 rpm (see Table 9.2)[2] Even moderately elevated temperatures can hasten the deterioration of internal components (valve seats and surfaces) and shorten the effective performance life of lubricating oil.

Water-cooled engines, in which temperatures are controlled with a jacket of fluid coolant around the cylinders and combustion chambers, have thermostats that regulate heat dissipation through a radiator. When burning ethanol, the frequency and duration of the thermostat cycle is reduced owing to the fact that there's less heat to dissipate. In some cases the thermostat may have to be replaced to maintain coolant temperatures at an optimal level.

Table 9.2
Exhaust Gas Temperature Results (Peak EGT)

2100 RPM	Average EGT	Percent Decrease EGT	Delta T
100LL	1326.0	--	--
10 Percent	---------	data not available	----------
20 Percent	1302.5	.02	- 23.5
40 Percent	1239.5	.07	- 86.5
60 Percent	1214.5	.08	- 111.5
80 Percent	1200.4	.09	- 190.9
90 Percent	1135.1	.14	- 190.9
E-95	1164.5	.12	- 161.5

Note: Data taken from "Flight Performance Testing of Ethanol/100LL Fuel Blends During Cruise Flight," page 45, Timothy J. Compton, M.S., Baylor Institute for Air Science. E-95 is a blend of 95 percent ethanol and 5 percent gasoline.

Emissions

The combustion of fuel in an engine generates by-products that we all know as emissions. The four main automobile emissions are hydrocarbons, carbon monoxide, oxides of nitrogen, and carbon dioxide (though others, such as particulates and formaldehyde, are also produced). It is clear that anthropogenic, or human-induced, pollutants are spiraling out of control and must be managed for the sake of our environment and our future well-being. But even from a basic economic viewpoint, a clean-burning fuel is a more efficient fuel. It reduces energy waste and extends the life of internal components by decreasing the incidence of damaging carbon residue within the engine.

Gasoline, as a compound hydrocarbon, is not a particularly clean-burning fuel. Every motorist knows the danger of allowing a car to idle in an enclosed space; what may be lesser-known is that auto emissions are responsible for 53 percent of the typical US family's annual contribution of CO_2 — nearly 27,000 pounds per year.[3]

Ethanol, in comparison, burns nearly pollution-free. It already contains oxygen integral with the fuel, which can lead to a more homogenous combustion.[4] It burns with a faster flame speed than gasoline, and it does not contain additional elements such as sulfur and phosphorus. All these factors work in ethanol's favor with regard to emissions — as well as the fact that its lower exhaust gas temperatures tend to reduce NO_X (nitrogen oxide) emissions specifically. At low load, NO_X emissions with ethanol are lower than with gasoline, due to differences in the latent heat of vaporization and in the combustion speed. The low exhaust temperatures produced when using ethanol enable a balanced combustion mixture along the whole operational range, and make it possible to use a 3-way catalyst to reduce NO_X emissions, even at full load.[5]

Coming up with accurate emissions data for a "typical" engine is difficult, however, for a number of reasons. Tests have been conducted by a number of universities, the US Department of Energy's Solar Energy Research Institute (SERI) and its Office of Alcohol Fuels (OAF), the Environmental Protection Agency, various automobile manufacturers, and by the petroleum industry itself. The research spans many years, but the tests were conducted with diverse goals. Some concentrated on alcohol gasoline blends; others analyzed the use of methanol; many made modifications to the engines that altered results; some concentrated on engine wear patterns and longevity.

One program conducted in 1976 by aeronautical engineer Richard F. Blaser focused on retrofitting a 10-horsepower two-cylinder generator engine as a dual-fuel (ethanol-gasoline) powerplant for farm equipment adaptation. The modifications included an increase in compression ratio to 9:1, the installation of a wedge-shaped cap at the top of the piston, a re-machined cylinder head and combustion chamber, and the installation of an alcohol fuel feed and air-entraining fitting. During a variable load test, carbon monoxide levels were reduced from a 6 to 9 percent range to a 0.2 to 0.7 percent range. There was also a reduction in the quantity of unburned hydrocarbons from 150-200 ppm to 50-100 ppm at rated output. This represents an improvement of 96 to 98 percent for CO (carbon monoxide) and 50 percent for UHC (unburned hydrocarbons).[6]

Our own independent testing at *Mother Earth News* in June of 1979 involved the simple conversion of a 1978 Chevrolet Impala taxicab to straight ethanol for an impromptu press conference, at the height of what became known as the "second gas crisis." In that test, the vehicle (which was subject to stringent New York City pollution standards and had to pass four scheduled EPA emissions tests per year) registered 1.5 percent CO and 200 ppm HC exhaust content. After removing all pollution-control equipment from the engine except for the positive crankcase ventilation valve, we retested with results that astonished even the characteristically cynical New York cab drivers. Carbon monoxide levels dropped to 0.08 percent, with a corresponding drop in hydrocarbon emissions to 25 ppm. Together, emissions were reduced by over 90 percent using straight ethanol in a gasoline engine modified only with carburetor re-jetting and ignition timing advance.

It's probably safe to say that a decrease in CO and HC emissions can be expected with

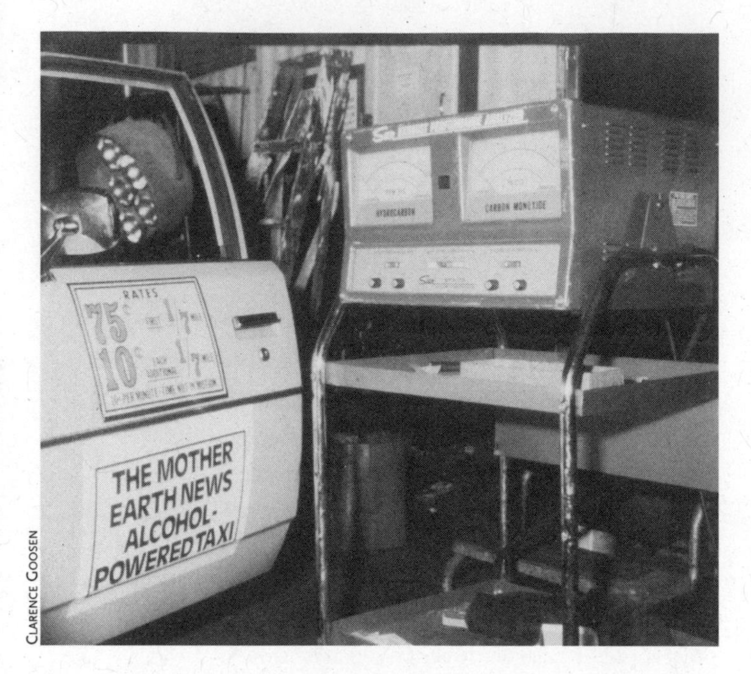

Fig. 9.2: *Carbon Monoxide and Hydrocarbon emissions testing on a New York City taxicab for a press conference in June, 1979.*

ethanol fuel, but the reduction level varies according to the vehicle type and technology used. Nitrogen oxide emissions are variable, depending on the factors mentioned above and the load placed on the engine. Exhaust-gas recirculation and catalytic converters should improve outcome significantly. The one unresolved issue is that of aldehyde emissions, which make up a small portion of the overall hydrocarbons. Even though they are statistically slight, aldehydes are chemically reactive and contribute to eye-irritating smog. The research that's been done on neat ethanol (straight ethanol with the required 2 to 5 percent denaturant added) shows an increase in aldehyde emissions four times that of gasoline. Still, the contribution is very small, and exhaust catalysts are designed to reduce it even further.

Performance and Fuel Economy

Fuel economy, or gas mileage, is relative to engine load and the mixture of air and fuel that's burned in the cylinders. Fuel, no matter what kind, must be mixed with air before it will combust completely inside an engine. Whether the fuel is blended with air before it enters the intake manifold (by means of a carburetor) or whether it is mixed in the manifold or directly in the combustion chamber (by means of fuel injection), a certain ratio of air to fuel must be attained in order for the mix to burn efficiently.

Each fuel has its own proportion, known as the stoichiometric ratio. Gasoline needs 14.7 parts of air to each part fuel. Ethanol, because it contains about 30 percent oxygen by weight, only requires 9 parts air to each part alcohol. These figures are theoretically precise, but in the real world, they will vary somewhat depending upon load and speed demands placed upon the engine.

A lean condition, in which the balance of air to fuel is greater than normal, results in better fuel economy, but higher exhaust gas temperatures. As more air (or less fuel) is introduced to the mixture, the engine may temporarily increase rpm, but performance will deteriorate, and if the leaning trend continues, damage to the valve facings will occur as internal temperatures reach a critical point. Conversely, a rich condition is one in which the balance of fuel to air is greater than normal. Adding more fuel (or less air) to the mix lowers exhaust gas temperatures and increases power output, but also lowers fuel economy and contributes to hydrocarbon emissions.

Most every contemporary automobile is equipped with fuel injection, which automatically compensates for load and environmental variables through a system of sensors and computerized controls. Cars, of course, are optimized for gasoline. But piston-driven aircraft can provide some insight into the performance characteristics of ethanol fuel because they operate in varying altitudes and weather conditions and are regularly exposed to extreme changes in air pressure and temperature. In aircraft, the pilot manually adjusts the air-fuel mixture to maintain an ideal ratio while atmospheric pressure and temperatures vary.

In certification testing done on a Lycoming O-235 aircraft engine in 1992, the maximum horsepower achieved by the engine when fueled by 100-octane Low Lead avgas (aviation gas) was approximately 125 horsepower. A 20 percent increase (25 hp) was gained when the fuel was changed to ethanol.[7]

Vehicles with fixed-jet carburetors or simple air-cooled engines don't have the flexibility to optimize ethanol fuel without modifications. But even these fundamental engines can be made to perform well on ethanol with a few basic changes, especially when ethanol is the dedicated fuel. What follows points up a few areas for consideration when contemplating a conversion to ethanol fuel.

Cold-Weather Starting

Probably the single biggest issue facing ethanol-fuel users is that engines have difficulty starting in temperatures below about 35°F to 50°F. This is actually not a major issue for reasons we'll see in a moment, but first let's look at why this occurs. The causes are related to ethanol's flash point and latent heat of vaporization, which we looked at earlier. Both are significantly higher for ethanol than they are for gasoline, so in effect alcohol fuel is less volatile, which can induce starting difficulties in cold conditions. Even gasoline itself isn't immune to this phenomenon, and in fact pump gas is "blended" for winter use by adding more volatile substances such as methylbutane in cold climates.

The common resolution to this problem is to start the vehicle on something other than ethanol — usually gasoline, though other, more aromatic fuels such as ether and propane can also be used. The booster fuel doesn't need to stay in the system long, just enough to kick the engine over. Once the engine is running, there's enough heat generated to vaporize ethanol sufficiently until it reaches operating temperatures, at which point the vehicle runs normally. Slightly more sophisticated is a system in which the ethanol is preheated at the carburetor or manifold by an electric element, which eliminates the need for a separate starting fuel.

Before moving on, it's worth mentioning that the E-85 ethanol fuel blend sold at service station pumps — 85 percent gasoline and 15 percent ethanol — has enough gasoline in it to start an engine in cold weather without the help of cold-starting aids. Manufacturer-built Flex Fuel vehicles (FFVs, for which the fuel was developed) and converted vehicles alike can use E-85 without using cold-starting systems.

Miscibility

The question of ethanol's ability to blend with gasoline is one that's unfairly plagued ethanol-fuel proponents for years. There is a concern that if the alcohol isn't anhydrous, or completely free of water, the ethanol will separate from the gasoline and stratify the fuel into two different elements. This worry is unfounded for a number of reasons. First of all, an engine dedicated to ethanol won't require any blending with gasoline. Petroleum fuels, if used at all, will remain in the cold-starting system, apart from the main supply of ethanol fuel. Furthermore, the vast majority of small-scale producers won't be making anhydrous ethanol in any event. The best we can achieve without a major investment in equipment is 190- or 192-proof ethanol, which has proven more than adequate for engine fuel.

Even in a situation where fuel blending is necessary, the water present in a good grade of fuel ethanol is neither a freezing hazard nor a separation agent. Pure alcohol is by nature hygroscopic, and after a period in either a storage tank or in the fuel tank of your car, it will absorb some water from the atmosphere or through condensation. This, in fact, is how Dry Gas,[R] the common fuel demoisturizing additive, works. It is pure alcohol, and it absorbs the molecules of water in your gasoline tank and dissolves them uniformly into the gasoline and alcohol blend, where they won't separate and clog the fuel filter or flood the fuel line. In essence, gasoline and water do not mix, but gasoline, water and ethanol are entirely miscible.

Low-proof alcohol (in the 160-proof range) continues to mix effectively with gasoline at room temperature and above. As the temperature drops, higher-proof alcohol is needed to establish a stable mix with gasoline. 190-proof ethanol will form a consistent blend at temperatures well below freezing.

Corrosion and Degradation

Certainly, the ability of ethanol to corrode metal parts and degrade soft components such as fuel lines, seals, diaphragms and so forth is a legitimate concern. Fortunately, the corrosive effects of ethanol are related to its water content. Later in this chapter, we'll look at some specific examples of component degradation and how to deal with it, but for now, there are two important points to keep in mind. First, low-proof ethanol is the real culprit when it comes to corrosion. Though it's true that many engines — particularly carbureted types — will run on 160-proof ethanol, their components are still prone to deterioration over time. The movement of ions in the water carries a current that's capable of slowly dissolving metals such as aluminum alloys or zinc. These problems effectively go away when the water content in ethanol falls below 5 percent, the equivalent of 190-proof. Secondly, much of the reputation for corrosion that alcohol has gained over the years is erroneously attributed to ethanol when it is *methanol*, the toxic race fuel and octane-enhancer, that is notoriously corrosive.

In the early 1980s, in response to the 1978 EPA approval for the use of gasohol (10 percent ethanol/90 percent unleaded gasoline) in the general automotive population, car manufacturers began making auto and truck components fully compatible with ethanol fuels. This included fuel tanks, fuel lines, pumps, sensors and carburetor/injector components. Over the course of 30 some-odd years, fuel blends containing between 6 and 15 percent ethanol have been sold in the US without significant repercussions. It's estimated that about 50 percent of America's gasoline now includes some amount of ethanol.[8]

Critics may point out that the ethanol blended into our gasoline is anhydrous, and contains only trace amounts of water. But in my travels to Sao Paulo and other cities in Brazil, ethanol-only vehicles routinely operated on hydrous or "hydrated" alcohol containing 4 or 5 percent water with no evidence of fuel-tank corrosion or component deterioration at all.

Now those of us with vehicles older than 25 years (and make no mistake — there are still millions of "veteran" cars, trucks, motorcycles, tractors, and other sorts of gasoline-powered equipment still in everyday use) will have decisions to make if we use ethanol fuel. Fuel tanks — even metal ones — are generally safe with high-proof ethanol, though some intake filter assemblies from later years may not be alcohol-resistant. If there's doubt, tanks can be cleaned and etched, then lined with a liquid tank sealer, which leaves a thin alcohol-impervious coating once cured. This is a do-it-yourself process available in package form through automotive restoration suppliers such as The Eastwood Co. and J.C. Whitney Co. (see Appendix B, "Resources and Suppliers).

Fuel pumps in older vehicles may contain some flexible components that will soften or swell in the presence of ethanol, but they can be replaced with aftermarket pumps fitted with modern neoprene. Very old fuel pumps may have varnished cloth diaphragms, which will disintegrate in alcohol. Fortunately, these pumps are made to be disassembled (oh, for the good old days!), and the diaphragms can be traced and re-cut from a sheet of elastomer or neoprene, then reinstalled.

Fuel lines are a mixed bag. Typical neoprene lines and hoses are ethanol-compatible, but clear plastic lines are likely to soften and swell or collapse — though not all types. Still, it's best to err on the side of caution and replace those clear hoses with modern neoprene or steel fuel lines.

Metallic components made of brass or stainless steel are unaffected by alcohol, and soft parts made of neoprene or Viton (a tradename fluoroelastomer) are likewise relatively safe. Gaskets are for the most part unaffected by ethanol.

Engine Modifications

In this section, I'll get into the details of converting a gasoline engine to alcohol fuel, and specifically the modifications and adaptations necessary to get the job done. I'm including engine principles in the segment following, particularly to support those new to mechanical applications. Thereafter I'll be addressing key components — carburetion, fuel injection, ignition timing, and cold-starting systems — followed by some specific modifications and adjustments that will make things easier for engines to accommodate the new fuel.

Principles of Engine Operation

By and large, the most common type of engine used in everything from lawn mowers to passenger cars is the spark-ignition internal

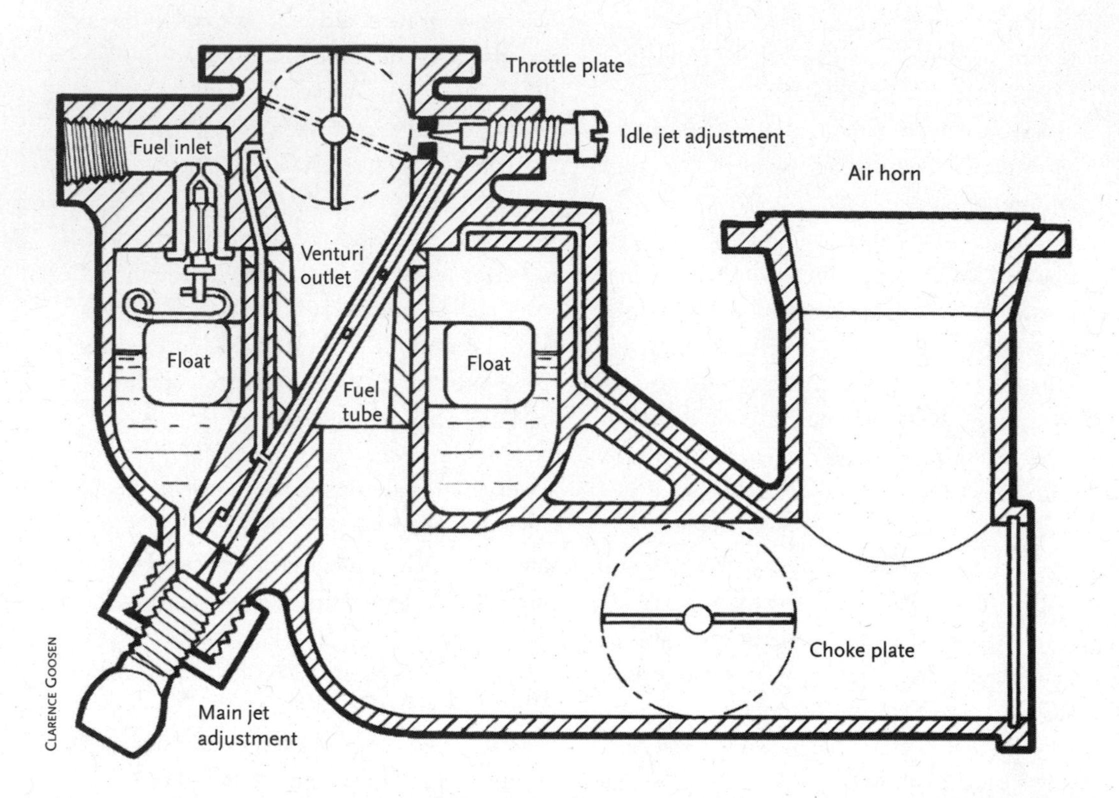

Fig. 9.3: *A schematic of a simple single barrel carburetor shows basic carburetor components.*

combustion engine (ICE). *Spark ignition* because the fuel is ignited by an electrical spark plug (as opposed to the compression ignition used in diesel engines), and *internal combustion* because the fuel burns inside the engine rather than external to it, as in a steam engine. These kinds of engines are known as reciprocating or piston engines, referring to the up-and-down motion of a piston in a cylinder.

The most common combustion sequence used is the four-stroke cycle, which has four stages: (1) the intake stroke, in which a mixture of fuel and air is drawn into the cylinder through a valve as the piston moves downward; (2) the compression stroke, in which the piston moves upward, compressing the fuel-air charge; (3) the power stroke, which occurs when the spark plug ignites the charge and drives the piston downward; and (4) the exhaust stroke, which forces combusted gases from the cylinder through another valve when the piston moves upward to complete the cycle.

The piston is connected to a crankshaft, which converts this up-and-down motion to the rotary motion of a shaft. Since the piston makes one up-and-down trip for each full rotation of the crank, one complete four-stroke cycle yields two crankshaft revolutions. So, an engine turning at 2,000 revolutions

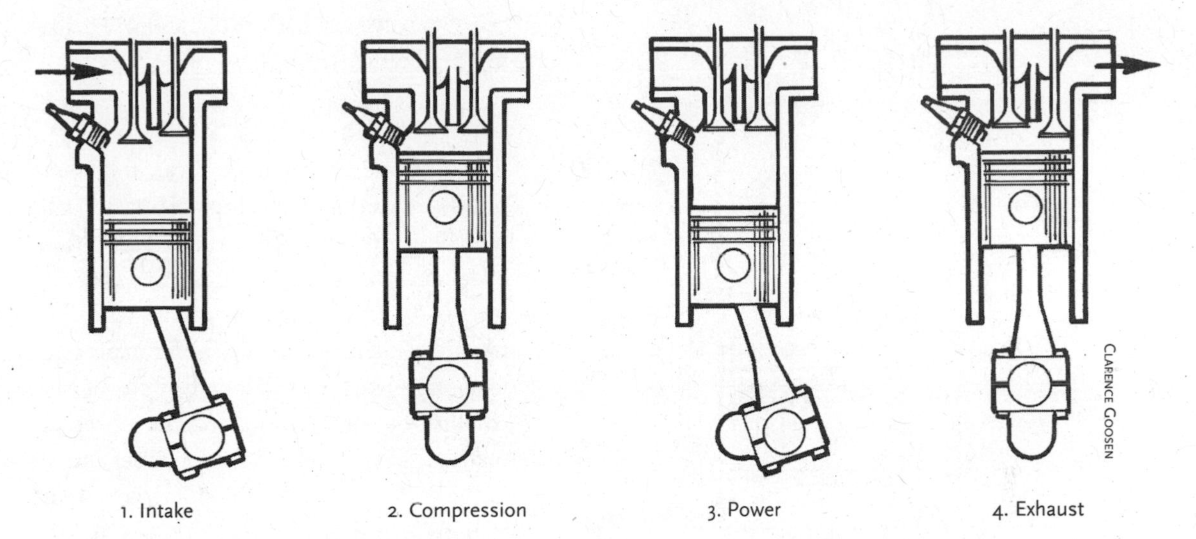

Fig. 9.4: *The four-stroke cycle.*

per minute completes 1,000 such cycles each minute.

Another, less familiar type of combustion cycle is the two-stroke cycle, commonly used in chainsaws, weedcutters, small motorcycles, and other applications where maximum power with minimum weight is essential. Two-stroke engines use ports instead of valves, which require a timed mechanical connection to operate them. These ports are controlled directly by the engine piston, and the engine crankcase (the chamber beneath the piston that houses the crank mechanism) serves as a compression pump and fuel-mixing chamber. A transfer port allows fuel to move from the crankcase to the combustion chamber, where the spark plug is located. The design is simpler and uses fewer components, making the engine lighter with no loss of power.

In the two-stroke cycle, when the piston is moving downward the exhaust port is opened first, allowing combusted gases from the previous cycle to escape. Shortly thereafter, the piston reaches the bottom of its travel (a position referred to as bottom dead center, or BDC), and the pressurized fuel mixture is forced through the transfer port into the combustion chamber, forcing out, or scavenging, any remaining exhaust fumes.

The second part of the cycle occurs when the piston moves upward, first closing the transfer port, then the exhaust port. The fuel mixture is compressed while the rising piston creates a vacuum that draws a fresh fuel charge into the crankcase through the intake port. When the piston reaches the top of its travel (top dead center, or TDC) the spark plug ignites the fuel mixture, driving the piston downward and initiating the cycle all over again.

In both four- and two-stroke engines, fuel and air is mixed precisely in a carburetor before being sent to the combustion chamber(s); engines with multiple cylinders use an

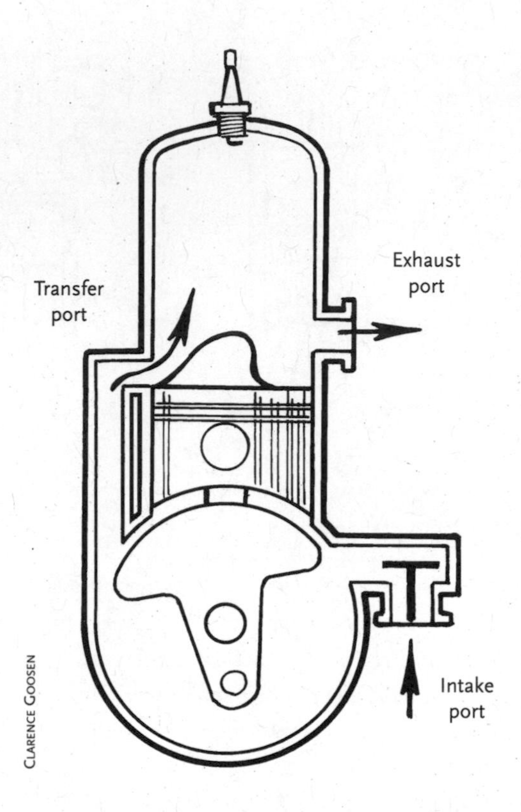

Fig. 9.5: *The two-stroke cycle.*

intake manifold to deliver the mixture in equal proportion to each cylinder. (In most contemporary engines, the carburetor has been replaced with a more efficient fuel injection system, which I'll discuss separately further on in this chapter.)

The rate of spark plug firing is precisely timed to the speed of the engine in relation to the position of the piston(s). This timing can be adjusted within a moderate range to suit fuel and varying driving conditions such as increased load or altitude changes. The term "ignition timing" refers to the point of spark ignition on a 360-degree scale corresponding to the rotation of the crankshaft. This point can typically vary between 5° and 45° before top dead center (BTDC).

Carburetion

Pop the hood of any modern car, and you're not likely to find a carburetor. But don't be misled — millions of cars, trucks, motorcycles, boats, aircraft, mowers, pumps, generators and tractors that were manufactured with carburetors are still in use and will be for many years to come. Since the earliest days of the internal combustion engine, carburetors have been the mechanism of choice for mixing fuel and air in the appropriate ratio for efficient combustion. As mentioned above, for gasoline, that ratio is about 14.7 parts air to 1 part fuel; for ethanol, the ratio is closer to 9 to 1.

A carburetor includes a small reservoir, or bowl, where liquid fuel pumped from the tank is maintained at a more or less consistent level by an internal float connected to a needle valve. This micro fuel supply feeds several circuits in the carburetor: an idle circuit that allows the engine to run smoothly at minimum rpm, a main circuit that allows cruising and high-speed operation, and, in more sophisticated carburetors, an accelerator pump and a power valve circuit that help in the transition from low to full power. Let's look at each individually.

Main Circuit

The main metering circuit is the *paterfamilias* of the carburetor. It may be an overstatement to say that all things flow from it, but not by much. The main circuit is responsible for controlling the fuel mixture in every cruising

How a Venturi Works

A carburetor is really just a glorified length of pipe with holes in it. But to get that pipe to answer the demands of an engine, it has to be modified with the addition of a *venturi*, a narrowed-down passageway within the pipe. This restriction is designed with smooth, tapered walls so incoming air is forced to pass through very rapidly, increasing its velocity by a factor of five or more. The increase in velocity creates a vacuum at the center of the venturi, where a strategically placed fuel inlet admits fuel, pushed from the float bowl by atmospheric pressure. When the liquid fuel meets the rush of incoming air, it is atomized into tiny fuel droplet particles, mixing thoroughly with the air before being ignited in the combustion chamber.

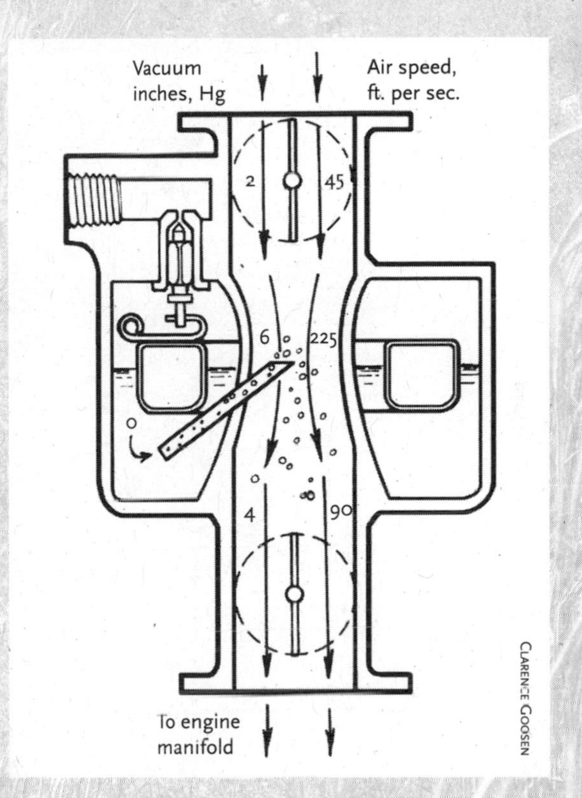

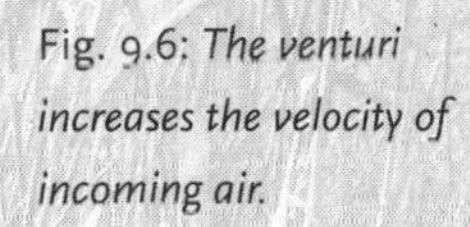
Fig. 9.6: *The venturi increases the velocity of incoming air.*

range save for idle and acceleration, but even then it still makes some contribution. Its job is to meter the amount of fuel entering the venturi area so it can be matched to the needs of the engine at any particular speed, and it does this through the use of a main jet.

In most carburetors, the main jet is a threaded brass plug with a hole of specific diameter drilled through the center of it. The size of the hole determines its area, which in turn determines how much fuel is allowed to pass through. Table 9.3 shows the relationship between the diameter of the hole and its area in square inches. These jet orifices are quite small: to give an example, the openings in the two-barrel carburetor of my 1980 Ford pickup were .042", about the thickness of the lead in a mechanical pencil. The smaller the opening, the less fuel that gets by. Alcohol requires more fuel flow, so the orifice needs to be enlarged by 20 to 40 percent in total area. In practice, somewhere between 30 and 35 percent is typical, but only experimentation and performance will tell.

The surgery can be accomplished with a power drill, but since brass is relatively soft, a pin vise is a better choice, since it's infinitely more controllable. It is a small, screwdriver-like handle that grips a drill bit and allows you to bore a clean hole by spinning the tool in your fingers. This is more jewelers' work than construction work, and there's far less opportunity

Table 9.3
Orifice Diameter vs. Area in Square Inches

Diameter (in.)	Area (sq. in.)	Drill No.	Diameter (in.)	Area (sq. in.)	Drill No.	Diameter (in.)	Area (sq. in.)	Drill No.
.0135	.000143	80	.0595	.002779	53	.1470	.016972	26
.0145	.000165	79	.0635	.003165	52	.1495	.017544	25
.016	.000200	78	.067	.003526	51	.1520	.018146	24
.018	.000254	77	.070	.003848	50	.1540	.018627	23
.020	.000314	76	.073	.004185	49	.1570	.019359	22
.021	.000346	75	.076	.004356	48	.1590	.019856	21
.0225	.000397	74	.0785	.004837	47	.1610	.020358	20
.024	.000452	73	.0810	.005133	46	.1660	.021904	19
.025	.000491	72	.0820	.005281	45	.1695	.022553	18
.026	.000531	71	.0860	.005809	44	.1730	.023506	17
.028	.000616	70	.0890	.006221	43	.1770	.024606	16
.0292	.000669	69	.0935	.006862	42	.1800	.025447	15
.031	.000755	68	.0960	.007238	41	.1820	.026015	14
.032	.000804	67	.0980	.007543	40	.1850	.026880	13
.033	.000855	66	.0995	.007771	39	.1890	.028055	12
.035	.000962	65	.1015	.008087	38	.1910	.028652	11
.036	.001017	64	.1040	.008495	37	.1935	.029392	10
.037	.001075	63	.1065	.008904	36	.1960	.030156	9
.038	.001134	62	.1100	.009503	35	.1990	.031086	8
.039	.001195	61	.1110	.009677	34	.2010	.031714	7
.040	.001257	60	.1130	.010029	33	.2040	.032668	6
.041	.001320	59	.1160	.010568	32	.2055	.033151	5
.042	.001385	58	.1200	.011310	31	.2090	.034289	4
.043	.001452	57	.1285	.012962	30	.2130	.035614	3
.0465	.001697	56	.1360	.014527	29	.2210	.038340	2
.052	.002124	55	.1405	.015496	28	.2280	.040807	1
.055	.002376	54	.1440	.016277	27	Note: The diameters selected correspond to standard number twist drill diameters.		

for damage with a hand tool. Numbered drill bits are organized in a numeric range to represent a variety of wire-gauge diameters in thousands of an inch, from size No. 80 (.0135") to size No. 1 (.228"), as shown in Table 9.4. The orifice diameters generally fall in the middle of this range; a No. 54 (.055") drill is a typical diameter for the jet of a single-barrel carburetor on a six-cylinder engine of around 3.8 liters displacement — about 230 cubic inches. Referring to Table 9.3, a 35 percent increase in area comes closest to a diameter of .064" — the approximate size of a No. 52 drill, and a diameter increase of 16.4 percent.

A Question of Percentage: Jet Diameter versus Jet Area

There's a significant difference between a 20 percent enlargement in the diameter of an opening and a 20 percent enlargement in its area. In short, increasing the area is a simple linear progression, while enlarging the diameter brings about exponential increase. So, augmenting the area of a No. 54 jet (.00237 square inches) by 20 percent gives .00284 square inches, only changing the jet size to No. 53 (.060" diameter). Increasing the diameter of a No. 54 (.0550") jet by the same percentage yields a .066" diameter, boosting the jet size to No. 51. As the percentage of enlargement increases, the disparity can account for a range of more than just a few jet sizes.

Curiously enough, the disparity doesn't seem to have catastrophic effects in practice. The Bernoulli theorem explains some of it, and though I won't even try to clarify the equation, I will say that it's based on the ratio of air mass to fuel mass and its relationship to flow rate, and that ethanol has a higher density than gasoline. I myself prefer an empirical measure of experimentation and results, which works well for most people, especially if they have analytical tools at their disposal. An exhaust gas temperature (EGT) gauge, reading from a sensor installed in the manifold close to the exhaust ports, will determine the leanest safe air-fuel mixture. An EGT over 1,350°F is excessive; progressively lower temperatures indicate a richer mixture, and the idea is to come to the leanest point without sacrificing performance. (EGT will read lower on ethanol fuel than on gasoline.) An exhaust gas analyzer, reading from a sensor clipped to the tailpipe, indicates carbon monoxide emissions, which should be close to zero with alcohol fuel. If the level rises while jetting progressively larger for performance, the jet opening is likely oversized. That condition will waste fuel and elevate nitrogen oxide levels.

With more complex carburetors, the relationship between main jetting, the power valve, and the accelerator pump becomes somewhat less obvious without test apparatus. If the power valve opens prematurely or is calibrated at a high delivery rate, larger jetting doesn't have as significant an effect, because the valve is providing additional fuel. Likewise, the accelerator pump will deliver a significant measure of fuel in transition from low to high rpm. Still, at steady cruising speeds, if the main jet is too small, the lean condition will manifest itself in performance and with evidence of overheated spark plugs — the electrodes and insulator will have a glassy, light-colored appearance, with possible ceramic cracking. Too large a jet is more difficult to discern, which is why testing should be done with each jet-size change.

I can tell you from experience that it's best to remove the carburetor from the engine before you do any work on it, both to make access easier and to eliminate the risk of getting metal chips in the manifold. Remove the air filter housing and all connecting hoses, then disconnect the throttle linkage and the choke linkage, if it's not self-contained on the body. Remove the fuel line and any vacuum hoses. Unthread the stud nuts or remove the screws holding the carburetor to the manifold.

To expose the main jet(s), you'll need to remove the air horn, or the top of the carburetor, by first extracting the screws that hold it in place. If you can access a Chilton or Mitchell auto repair manual for your particular make and year model, the carburetor exploded diagrams will help immensely; otherwise you can

Table 9.4
Drill Size Diameters for Fractional and Number Twist Drill Sets

Drill No.	Diameter (in.)	Drill No.	Diameter (in.)	Drill No.	Diameter (in.)	Drill No.	Diameter (in.)
80	.0135	58	.0420	38	.1015	18	.1695
79	.0145	57	.0430	37	.1040	11/64	.1719
1/64	.0156	56	.0465	36	.1065	17	.1730
78	.0160	3/64	.0469	7/64	.1094	16	.1770
77	.0180	55	.0520	35	.1100	15	.1800
76	.0200	54	.0550	34	.1110	14	.1820
75	.0210	53	.0595	33	.1130	13	.1850
74	.0225	1/16	.0625	32	.1160	3/16	.1875
73	.0240	52	.0635	31	.1200	12	.1890
72	.0250	51	.0670	1/8	.1250	11	.1910
71	.0260	50	.0700	30	.1285	10	.1935
70	.0280	49	.0730	29	.1360	9	.1960
69	.0292	48	.0760	28	.1405	8	.1990
68	.0310	5/64	.0781	9/64	.1406	7	.2010
1/32	.0312	47	.0785	27	.1440	13/64	.2031
67	.0320	46	.0810	26	.1470	6	.2040
66	.0330	45	.0820	25	.1495	5	.2055
65	.0350	44	.0860	24	.1520	4	.2090
64	.0360	43	.0890	23	.1540	3	.2130
63	.0370	42	.0935	5/32	.1562	7/32	.2188
62	.0380	3/32	.0938	22	.1570	2	.2210
61	.0390	41	.0960	21	.1590	1	.2280
60	.0400	40	.0980	20	.1610		
59	.0410	39	.0995	19	.1660		

Fig. 9.7: *Removing the carburetor air horn fasteners.*

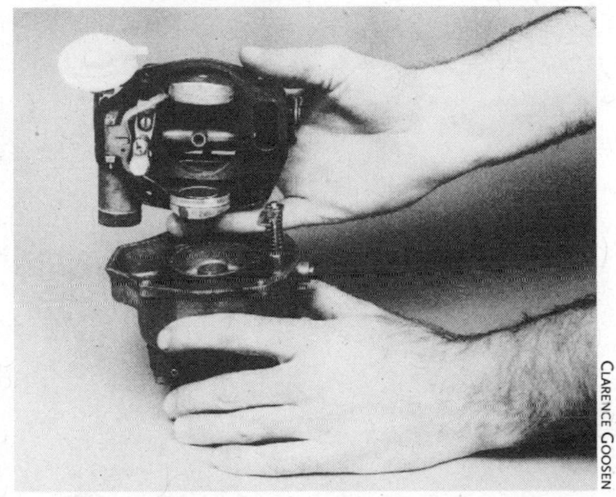

Fig. 9.8: *Lifting the air horn to expose the float bowl.*

buy a carburetor rebuild kit from an auto parts store, which will get you the diagrams you want plus the gaskets, O-rings, float valves, and other parts you might need to reassemble the unit. Drain residual fuel from the float bowl by inverting the carburetor body.

The main jet(s) are either threaded into the base of the float bowl or, more likely with single-barrel carburetors, mounted in a cored tower that's integral with the air horn and submerged in the bowl when the carburetor is assembled. Remove the jet (or both jets, in a two-barrel carb) and determine, with your selection of drills, what diameter orifice it has. A snug but not tight fit is ideal. Start out with a 15 percent diameter increase (this is equivalent to a 31 percent increase in area) and pick the closest number bit that meets the criteria. For example, if the jet was originally .055" in

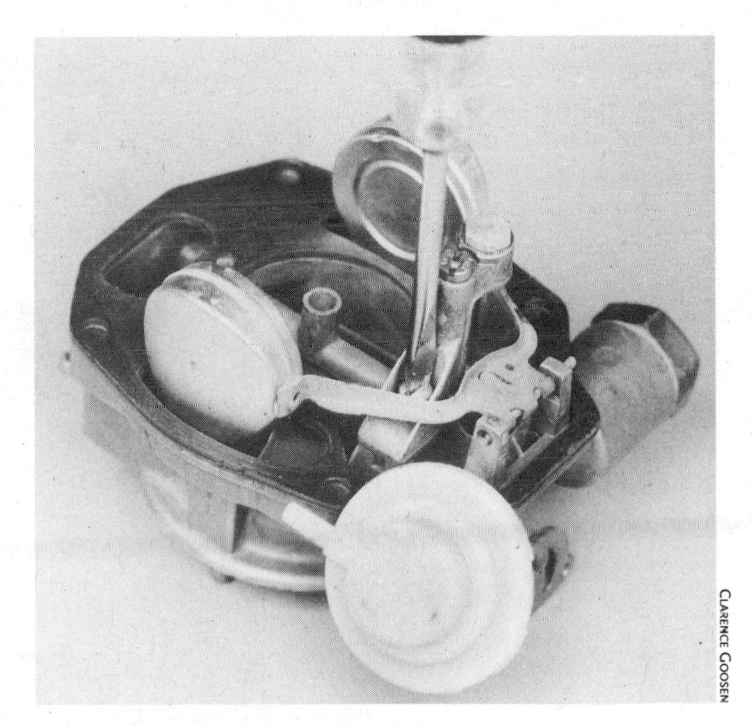

Fig. 9.9: *Removing the jet tower.*

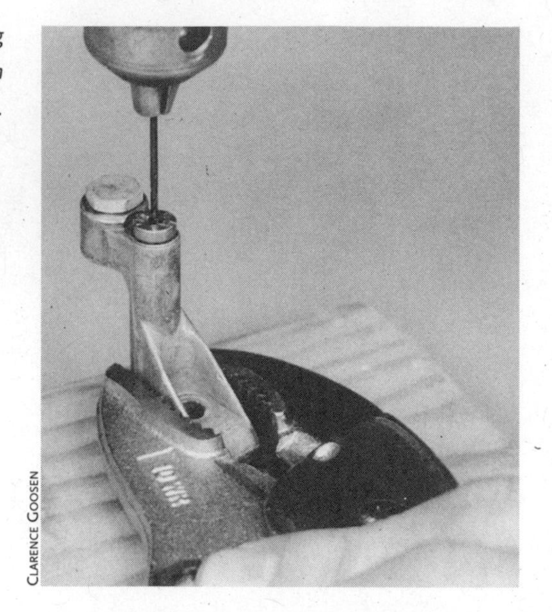
Clarence Goosen

Fig. 9.10: *Drilling the jet orifice with the proper size bit.*

diameter, multiply .055" by 1.15 to get .06325". The closest number bit to that is No. 52 at .0635".

Mount the bit in the pin vise and carefully bore out the opening, perpendicular to the flat surface of the jet head. Do not grip the bare threads of the jet with pliers. It may pay you to locate a damaged carburetor body and thread your jets into that when drilling for a secure mount. Also, if you can locate replacement jets of any size, it'd be a good idea to have them on hand for experimentation, since you may need to go up or down a drill size to achieve your goal. Jets are not always easy to come by, but you may be able to order them from a franchised dealer's parts department, an auto parts

Don Osby

Fig. 9.11: *Some carburetors use a metering rod and matching jet to control fuel flow more precisely than a jet alone would do.*

store, a specialty restoration supplier, or through a search on the Internet.

Some manufacturers do not employ fixed-orifice jets but instead use a tapered or stepped metering rod that moves up and down within a brass jet to vary the amount of fuel passing through the opening with changing throttle position. These jets can similarly be bored out, but to a lesser degree. A 10 to 25 percent increase range is more feasible here, or you can take the metering rod to a machine shop and have it turned down slightly to achieve the same effect.

After reassembling and mounting the carburetor, you can make a trial test (after first making modifications as needed to the idle circuit, described below) to see how close your estimates were. If the vehicle cruises well at steady highway speeds, but there's a strong odor of alcohol in the exhaust, the mixture is probably too rich. Do not expect it to accelerate hard or pull a steep hill without stumbling, because you haven't modified the accelerator pump or the power valve yet. If, on the other hand, the engine misfires and surges at cruising speeds, the mixture is lean. Either condition must be corrected, as the former will wash lubricating oil from the cylinder walls and the latter will burn spark plugs and valves, and create excessive levels of nitrogen oxide.

Accelerator Pump

The transition from idle to off-idle, and bringing the engine to speed, can be problematic even for a conventional gasoline-powered engine. The reason is that when you press the accelerator pedal down, the carburetor throat opens wide, dropping the manifold vacuum and reducing the effectiveness of fuel atomization. The fuel droplets that are introduced to the venturi area condense on the walls of the intake manifold and don't entirely make it to the combustion chamber, temporarily creating a false lean condition. The result is hesitation, stumbling, or even stalling in worst-case circumstances.

To alleviate this situation, carburetors are designed with a small pump linked to the throttle mechanism. The pump is fed from the float bowl and contains a one-shot charge of fuel that enters the venturi and is replenished on the return stroke. Primitive as it is, this approach gets the job done, albeit with some cost to fuel economy and emissions, since the air-fuel ratio is robustly enriched during the brief window. Nonetheless, what's good for the goose is good for the gander, and alcohol-fuel vehicles need the same juice boost during acceleration, just more of it, and usually for a longer duration.

If you're fortunate, your carburetor will have an adjustable or two-position setting on the pump lever, originally intended for winter/summer seasonal adjustment but perfect for ethanol purposes. The lever rod's inner setting, closest to the lever pivot, provides a short stroke and thus less fuel; the outer position, at the end of the lever, is the one to which you should move the rod to deliver more fuel volume.

It is possible to drill the accelerator pump orifice slightly larger — no more than 15 or 20 percent or so in area — to dispense more fuel, particularly on carburetors with internal pump assemblies. However, the opening is extremely small and often difficult to access.

DON OSBY

Fig. 9.12: The accelerator pump linkage is set on the "summer" or short stroke position. To lengthen the stroke and provide more fuel, the rod can be moved to the outer position, seen at the end of the lever.

On carburetors with removable discharge towers, the job is simplified but the openings are still small. One option is to replace the external pump assembly itself with a larger-capacity one that offers more duration (you'll have to match bolt patterns for fit), though there's no guarantee that this will solve the problem. And, you can always opt for a more accommodating carburetor. Manufacturers such as Holley and Weber make performance units to fit many engine applications that are markedly more adjustable than their factory-stock counterparts, and these companies carry a full line of replacement and modification parts as well.

However, here's something for the curiosity files: When colleague Ned Doyle of our old alcohol fuel crew converted his 1979 Harley Davidson Sportster to ethanol, he used the standard rules of thumb to enlarge his main jet and idle circuit orifice. But for some reason, the Keihin carburetor provided better off-idle response when he decreased the length of the accelerator pump stroke, possibly because it prevented fuel loading at a critical moment.

Power Valve

An often-overlooked component of the carburetor (and understandably so, since it's not visible from the outside) is the power valve, a spring-loaded control device that responds to low manifold vacuum. Low vacuum occurs when the throttle is open and the engine is under load, such as when climbing a steep grade. In this situation, the engine is still building rpm's, but the accelerator pump has already spent its charge. The power valve is normally pulled shut by the 16 to 20 In. Hg of vacuum pressure developed by air being sucked through and combusted in the engine; when that vacuum drops below 8 In. Hg or so, the power valve's spring lifts the valve door, triggering a flow of fuel to the venturi through an internal jet. It's a crude, but effective, resolution to the dilemma.

Power valves are replaceable in many automotive carburetors. If you can provide the model and series information from the metal ID tag on the carburetor, a parts house can search its catalog for a replacement valve with a higher vacuum release threshold, in the 12 In. Hg vicinity. Its greater spring tension will allow the valve to open sooner, delivering fuel in the critical period after the accelerator pump has given all it can. If you're the tinkering type, some power valves have spring keepers that allow them to be disassembled; you may be able locate a higher-tension spring to replace the original (make sure it's the same diameter as the original so it won't bind in the chamber). Another option is to fabricate a spacer collar from a small piece of copper tubing or a stack of stainless steel washers. Placed between the spring and its keeper, they will compress the coils and increase spring tension.

Idle Circuit

The idle circuit is a detour around the main circuit that allows a richer than normal fuel mixture for smooth idling. It's no more than a passageway connecting the fuel bowl with the carburetor throat, just below, or after, the throttle plate. When the throttle plate is closed, this circuit is the only one supplying fuel to the engine. A threaded needle valve, usually positioned at the orifice where the fuel enters the throat, controls the diameter of the opening to allow less or more fuel to enter, depending on how far the needle is threaded in or out.

For the engine to run optimally on alcohol fuel, the size of the opening needs to be increased by about 35 percent, or approximately the same increase in area that the main circuit requires. In some cases, simply backing out the needle screw will accomplish this goal. The best procedure here is to first very slowly turn the screw clockwise until the needle tip rests gently against its seat. Count the number of turns or half-turns it takes to get there so you can reset to the original starting point if needed. I can't emphasize enough how delicate the needle and seat can be. If you tighten the screw too hard or too quickly, it will score the tapered needle and ruin it, so be very careful.

At this point, you can start the engine and back the screw out incrementally until the idle begins to smooth out, then back the screw out one-half turn more. Backing it out further may not appreciably change the idle speed or improve its steadiness, but it will waste fuel. In some cases, the screw may be adjusted correctly, but is threaded out so far that it's in danger of vibrating loose. If that happens, shim the spring with a couple of small washers to maintain its original tension.

Ideally, the idle speed in rpm will be at or around the same level that it was on gasoline. More likely, with smaller engines especially, you'll have to kick up the rpm a bit to get a reliable idle. Picking up the rpm level *too* much, however, can invoke an intermediate fuel circuit in more complex carburetors, which defeats the purpose. Where that point is varies with each engine, but generally an idle speed of 1,500 rpm is probably too fast.

If backing out the screw completely still doesn't let enough fuel in to achieve a satisfactory idle, then the orifice itself will have to be enlarged. As with the main circuit, this is best accomplished with a pin vise and a numbered drill bit. Remove the carburetor from the engine and take out the needle screw altogether to expose the seat at the back of the threaded cavity. Determine the size of the existing orifice by testing it with different sized bits for a snug fit. Once that's established, choose a bit that's 25 percent larger, and use the pin vise or a small drill to carefully enlarge the hole. Take care not to damage the threads in the carburetor body. You'll notice that I said 25 percent and not 30 percent or 50 percent,

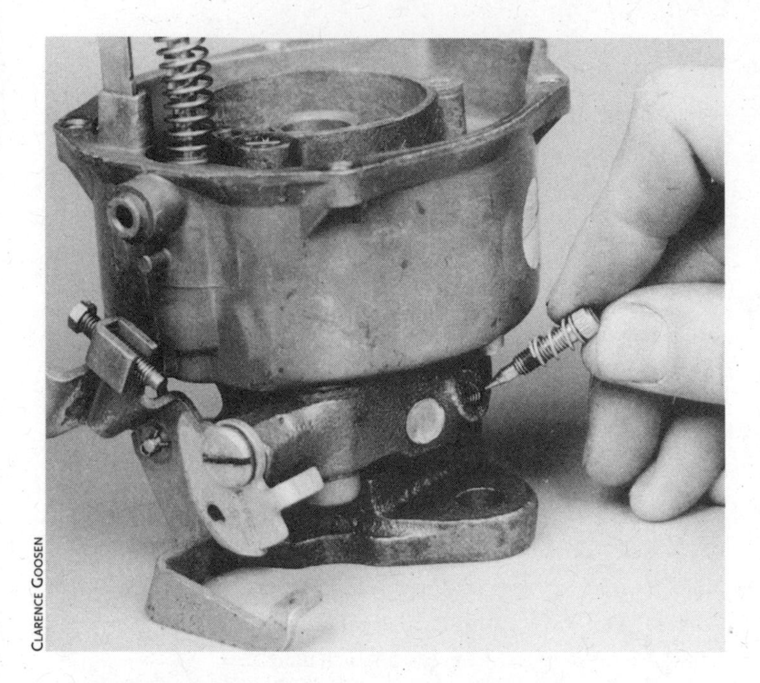

CLARENCE GOOSEN

Fig. 9.13: *Removing the idle screw exposes the orifice seat beneath it.*

CLARENCE GOOSEN

Fig. 9.14: *The orifice can be enlarged with a properly sized drill bit.*

as I've seen in some references. This is because the idle orifice is usually fixed and cannot be replaced. Starting out small will not do any harm; you can always go back to enlarge again if needed. More importantly, the run of the needle may well increase the area 20 or 25 percent on its own. Adding 50 percent on top of that would be excessive.

You should also be aware that some carburetors contain idle air bleeds to enhance movement of the air-fuel mixture. Increasing the diameter of the bleed orifice by up to 50 percent can improve the flow velocity for off-idle performance. Again, start with a small bit and work your way up in size.

There are several ways to determine if your adjustments are accurate. The first is by ear, which for the idle circuit may well be close enough because the engine's rpm is stable at idle. Turn the screw inward until the idle roughens, then back it out until the idle smoothes out. A fraction of additional turn outward will compensate for temperature variations and other incremental changes the engine might experience. Better yet, if you have access to a tachometer, you can use it to register the rpm levels while making your adjustments. While backing the screw out, seek out the point at which the revolutions per minute are highest, just after the engine has transitioned from a rough to a smooth idle.

Alas, there are many, many different carburetor designs and along with them diverse idle circuit configurations. It may take some detective work and experimentation to determine the flow within the circuit and how best to access the fuel orifice to enlarge it. B.V.

Alvarez, a brilliant research staffer at Mother Earth and top-notch mechanic in his own right, came upon the idea of using cigarette smoke to establish air vent and fuel paths inside the carburetor. It's not always easy to figure out which hole leads where without a schematic diagram, but smoke always finds its way. If you should come upon a carburetor with a fixed idle orifice allowing no adjustment (or one so convoluted as to make it nearly impossible to modify) you can always replace the carburetor. Be sure the replacement has a comparable flow rate (in cubic feet per minute, or cfm), the same manifold bolt pattern, and the linkage and fuel/vacuum fittings in the same place as the original.

Float Level Adjustment

The float within the carburetor float bowl governs the amount of fuel entering that small reservoir by controlling its flow. When the bowl is full, the float — which is hinged at one side — rises on the liquid's surface and shoves a small needle valve into a seat, shutting off incoming flow. As fuel is drawn into the venturi and the reservoir is emptied, the needle drops, allowing more fuel to enter. This cycle repeats itself dozens of times each hour the engine is running.

The level of fuel in the bowl is vital to proper fuel delivery. If it's too low, demanding acceleration may starve the engine of fuel before the bowl can refill. Too high and excess fuel can leak into the carburetor throat, wasting fuel and causing flooding. The factory shop manual or a carburetor rebuild kit will give the correct setting for the float height, usually measured from a reference point on the carburetor body. Since ethanol has a higher density than gasoline, the float will be slightly more buoyant in alcohol than it would be in gasoline, and the original settings will be incorrect. You'll have to readjust the height by bending the float arm tab to maintain the proper fluid level (not the float level) at the point where it would be for gasoline. The position of the tab merely lifts or lowers the needle closer to, or further from, its seat.

Floats themselves are made of brass, plastic, or occasionally a closed-cell foam material. Normally, none of these are affected by ethanol, but some older plastic floats can be degraded with extended exposure. Also, very old floats are made of cork painted with a shellac, which alcohol will dissolve. If you anticipate a problem, you may be able to find a brass replacement float for your particular carburetor. If not, you can seal the existing float with a thin layer of epoxy, which is alcohol-resistant. Don't overdo it, because too much may weight the float beyond your ability to correct with tab adjustments.

Chokes

A choke is a door or flap in the air horn of the carburetor that controls the amount of air allowed into the unit. When the choke is closed, it restricts air to the carburetor, enrichening the air-fuel mixture to aid in cold starting. In many vehicles, the choke is controlled by a bimetallic thermostatic spring that is either mounted in a well atop the intake manifold or attached to the side of the carburetor in a housing that picks up warm air from a tube routed to a shroud over the exhaust manifold.

In either case, as the engine warms up, its heat uncoils the spring, which slowly opens the choke through a linkage mechanism.

Since an alcohol-fueled engine takes a longer time to warm up, the choke often opens sooner than it should, leaning the air-fuel ratio at a point where it still needs to be rich. There is sufficient heat available at the manifolds to open the choke, but not enough to temper the engine block. With carb-mounted choke housings, you may be able to retard the opening sequence by loosening the three screws on the keeper ring surrounding the plastic cap and turning the housing to tighten the choke spring. Chokes that are electrically controlled may also have some adjustment range in them. But the only foolproof way to assure that the choke will be at the position you want it when you want it there is to install a manual control.

Manual control cable kits are readily available at auto parts stores and fasten to the existing choke linkage with a minimum of effort. The other end of the cable with the control knob is routed through the vehicle's firewall and mounted at a convenient spot beneath the dash. The cable knob is pulled to close the choke completely for starting, then it can be adjusted to a suitable partially open position until the engine gets to temperature, after which it can be fully opened.

Thermostat Replacement

The engine thermostat is a heat-controlled valve that holds the coolant within the engine block until the engine has reached its optimal operating temperature. Once that occurs, the valve opens, sending coolant to the radiator where it's cooled before returning to the engine block to be reheated. If your vehicle's heater isn't getting warm or the coolant is not being brought to its optimal temperature, you may have to replace the thermostat with one with a higher temperature rating. Several heat ranges are available, and a replacement in the 200°F range would be best, to bring temperatures back into the optimal range.

Changing the thermostat is a simple matter of draining the coolant and unbolting the thermostat housing from the engine block, at the end of the upper radiator hose. The thermostat itself rests in a seat that is cast into the housing, which is sealed with a gasket. Do not tape cardboard over the radiator as a substitute for replacing the thermostat; it may cause overheating at higher ambient temperatures. Use that technique only to boost coolant temperatures temporarily to determine whether a replacement thermostat is needed.

Fuel Preheating

This is a somewhat sensitive technique that can easily be overdone, but is quite effective if executed correctly. Carbureted engines especially have problems in atomizing fuel — including gasoline — to the extent necessary to assure truly efficient combustion. The best we can hope for is a happy medium, in which the fuel droplets are broken up into small enough particles to burn reasonably well. But even that occurs mostly during cruising or fixed-throttle operation, and suffers during idle, warmup and acceleration.

Warming the fuel before it enters the carburetor helps it to break apart into fine droplets more easily and discourages it from condensing into large droplets as it passes through the intake manifold. You may then wonder, if that's the case, why auto manufacturers don't routinely include fuel-warming circuits on their factory cars? The reason is because gasoline doesn't take to heating very well, and in fact is formulated to boil at temperatures as low as 90°F in cold climates. The increased oxygenation in modern fuels has lowered the boiling point of pump fuel significantly, to the point where vapor lock can be a problem in carbureted vehicles. Vapor lock, or boiling in the fuel lines, isn't a problem in fuel-injected engines, which can compensate for it and don't have to deal with fuel-atomization issues in any case because of injector pressures and the fact that the intake manifold handles incoming air and no fuel.

We know that ethanol boils at around 173°F, so it's safe to heat the fuel to near that point, using hot engine coolant to do the job. The simplest method is the one we used in the old Mother Earth days, which was to cut a section out of the upper radiator hose and replace it with a length of copper pipe sized to fit snugly inside the hose ends. Before installing it, we wrapped a few turns of copper tubing around the pipe and soldered it to the pipe in several spots. The free ends of the tube could then be spliced into the fuel line, after cutting out a section of the line. If you have access to a tubing bender, flared steel brake line can be substituted for the copper tubing so the connections can be made with threaded fittings rather than with neoprene hose and worm clamps.

Now here's the sensitive part: the number of turns wrapped around the pipe (and whether or not it is soldered, and how frequently) will determine how much heat is transferred to the fuel. It is a matter of trial and error, but any amount of heat up to about 160°F is an improvement. The downside to using this technique is that the coolant temperature in the upper radiator hose can rise above the boiling point just after the engine is shut down (remember that it's pressurized, so it won't boil over even at 212°F). That much heat can boil the fuel in the lines, forcing liquid alcohol past the needle valve on the carburetor float, flooding the engine manifold.

CLARENCE GOOSEN

Fig. 9.15: *A few wraps of copper tubing around a copper insert in the upper radiator hose serves as a simple but effective fuel preheater.*

To correct this, some enterprising souls have used coolant-to-fuel heat exchangers modeled after the Liebig condenser described in the previous chapter to maintain a more-or-less equal volume of fuel and coolant in the device. By plumbing into the heater supply hose (a smaller diameter hose that draws its coolant from the engine block) the heat transfer is more consistent than it is from the upper radiator hose (which is downstream from the thermostat) and somewhat isolated from the residual heat of the engine block.

With air-cooled engines on motorcycles, early Volkswagens, and most all power equipment, your heat source becomes the exhaust manifold or a portion of the exhaust pipe/muffler. Innovation is the key here, and you'll likely have to experiment with how much fuel line to expose to exhaust heat. One approach is to fabricate a sheet-metal shroud to cover the hot engine part, and then fasten the rerouted section of fuel line to the outside of that. Another option, used by the owner of an Indian motorcycle I saw at an alternative-fuel competition some years ago, was even cleaner. He passed the copper fuel line through a section of ½" copper pipe, with its ends bent at 90° and flattened. Holes were drilled at each end of the pipe to accommodate the line, and the flat ends were fastened beneath two cylinder-head bolts. Neoprene hose connected the copper line to the fuel supply and the carburetor. This, by the way, raises an important point: copper should not be fastened directly to fixed threaded fuel fittings because it will fatigue and crack with vibration. Use neoprene connectors and worm clamps, or go to all-steel fuel lines.

Electronic Feedback Carburetors

As early as 1975, auto manufacturers began to realize that the federal government could lay a heavy hand upon the business of making cars through its legislation. They had already seen the effects of advocacy and public opinion in matters of safety, in particular with the efforts of attorney Ralph Nader in targeting Chevrolet's 1960s Corvair in public hearings and through his book, *Unsafe at Any Speed*. This time the pressure was environmental, first with legislation to remove tetraethyl lead from motor fuel, then with ongoing efforts to reduce automotive emissions.

So, in a period beginning around 1979 and ending in 1993, when Mazda quit making its carbureted compact pickup truck, most manufacturers relied at some point on electronically controlled carburetors to meet more stringent emissions and fuel economy standards. These were called electronic "feedback" carburetors because they utilized information from electronic sensors — most importantly, an oxygen sensor in the exhaust manifold, along with temperature, airflow and rpm sensors — to adjust the air-fuel mixture according to engine conditions. An electronic control unit (ECU) or Microprocessor Control Unit (MCU), or simply a "black box," processed the information and fed it back to the carburetor for the changes.

Though the inclination may be to tear off the feedback carb and replace it with something fully mechanical, these transitional devices, if working correctly, can actually operate quite acceptably on alcohol. The trick is, not all of them have the broad range of adjustment needed for the richer ethanol air-fuel mixture. My 1986 six-cylinder Ford F-150 (4.9-Liter) reverted to an "open loop" condition when the black box was inadvertently damaged (meaning it ran at rich mode, with all the power you could ask for). Its 35 percent drop in fuel mileage suggests that it probably would've done just fine with ethanol if the ignition timing were adjusted to compensate

Dual-Fuel Carburetors

Here's an example of what you can do if you're clever at fabrication. The two-carburetor setup in the photo was installed on the 1977 *Mother Earth News* Chevrolet van we used to tour the country in promoting our alcohol fuel program. As a matter of convenience, it was designed as a dual-fuel system, so if the road crew should happen to run short of ethanol in unfamiliar territory, the vehicle could be transferred back to gasoline at the flip of a switch. Dual tanks and fuel lines were installed to accommodate both fuels.

The carburetors were side-draft Carter YH models commonly used in marine applications, mounted to a homemade "Siamese Y" manifold welded up from muffler tubing and exhaust flanges. The gasoline carb was left unmodified and came fairly close to matching the cfm flow of the 250 cubic-inch six-cylinder engine's original downdraft carburetor. The other carburetor had its main jet enlarged in diameter by 30 percent, its idle orifice enlarged (by 20 percent in diameter), and the idle transfer slot expanded to improve the off-idle transition. The idle circuit passageways were also bored out with a No. 25 drill to improve airflow. The float level was raised by 5/32-inch and the accelerator pump discharge diameter increased by 20 percent. To aid in acceleration from idle, the fuel wells that supplied the venturi tubes were also enlarged for greater capacity, and given booster air bleeds.

The unique part of this project was the dual-throttle linkage, which was made from two 4-inch door hinges. An anti-dieseling GM solenoid shifts the reworked hinge assembly to the right or left, depending on the fuel being used. Only the vacuum port on the alcohol carburetor was used for distributor vacuum advance; the other one was plugged, and the distributor static advance was increased by 12 degrees to accommodate both fuels moderately well. At idle, both carburetors drew equally, so the ratio of ethanol to gasoline was fully adjustable for the smoothest idle. No individual fuel line shutoffs were needed because the inactive carburetor saw no airflow at speed.

BACKHOME PHOTO COLLECTION

Fig. 9.16: *A dual fuel twin-carburetor setup fabricated for a 6-cylinder Chevrolet engine. One carburetor operates on unleaded regular gasoline and the other is optimized for alcohol. A specially designed solenoid-controlled linkage makes the switch between one or the other as needed..*

for the new fuel. Because feedback carburetors are so complex and inclined to going amiss, alcohol enthusiasts usually replace them as soon as they fail, either with conventional carburetors or with throttle body injector units (see below), which bolt right on to the intake manifold.

Fuel Injection

1990 was a watershed year of sorts in automotive history, because in that year the last carbureted American passenger car, a 5.0 Liter Oldsmobile V-8, left the Detroit assembly line. After years of struggling to meet federal emissions standards and fuel economy goals using cobbled-up plumbing and increasingly complex carburetors, Detroit finally acquiesced to the reality that fuel injection's time had come.

Mechanical pressure injectors actually appeared on diesel engines around 1930, and, during the Second World War, they were refined for use in high-performance gasoline-powered fighter aircraft to gain horsepower and shed the restraining effects of gravity on liquid fuel. In the 1950s, a few exotic sports cars and competition racers used injectors, but it wasn't until the late 1960s that they became part of regular production, most notably on certain Volkswagen models. The earliest continuous injectors provided additional fuel on demand, either by increasing the fuel flow rate to each injector through a distributor, or increasing the pressure within the fuel line. Each cylinder had its own injector mounted in the vicinity of the intake valve. Later, pulsed injectors were developed that used electric solenoids to open and close a valve on each injector. With these, fuel delivery was controlled by how long the valve was held open during each pulse cycle.

Pulsed fuel delivery has been simplified by the development of common-rail fuel lines — almost like a liquid manifold for fuel — to replace the separate lines to each injector. The pulse timing, and duration, or pulse width, is controlled by the same type of microprocessor control unit used to govern the feedback carburetors. The data collected by the MCU to determine the most appropriate pulse width for conditions includes the amount of oxygen in the exhaust stream, airflow, ambient temperature of the air, temperature of the coolant, and engine rpm, among other things. As engines and computer controls get more sophisticated, even more information is collected and processed to include altitude, preignition, engine timing, and coolant, air and exhaust gas temperatures, which are all used to manage various systems within the engine far more efficiently than they were even a few short years ago.

Now that we have the basics down, let's look at the two common electronic fuel injection (EFI) configurations.

Throttle Body Injection

Throttle body injection, or TBI, was developed as an economical approach that combined most of the benefits of fuel injection with the cost-saving measures of single-unit simplification. TBI systems use an injector housing mounted on top of the intake manifold in place of a carburetor. A throttle plate beneath the housing controls airflow into the manifold.

One or more fuel injectors mounted in the housing direct a spray of fuel under pressure into the manifold, and the pulses are controlled by the MCU as described earlier.

The TBI system atomizes fuel better than a carburetor can and reduces emissions in the bargain, but the fuel-air charge still has to pass through the manifold where it's likely to condense, at least to some degree. Throttle body injectors offer some opportunities for the mechanically inclined in situations where carburetors can't easily be converted. Aftermarket TBI kits are relatively inexpensive and are essentially bolt-on operations.

Multi-Port Injection

The multi-port injection systems are more common and work a bit better than TBI systems because they have individual injectors in the manifold that direct pressurized fuel to the back side of each intake valve. The manifold, then, remains dry and handles only air, eliminating the condensation, unequal distribution, and fuel puddling that plagues "wet" manifolds.

Multi-port injection costs a little more to manufacture because there are more components, including injectors, involved, but the gain in fuel economy and improved emissions makes the expense worthwhile.

Converting Fuel Injection Systems to Alcohol

Right off the bat, fuel injection is more intimidating than carburetion because it's not entirely mechanical. The average Joe or Josephine can't easily see how it works. Further, the electronic circuitry is not transparent, and the snarl of cables and sensors is both confusing and scaled at a micro-level. Fortunately, the conversion options available do include mechanical modifications, and equally alluring is the fact that an entire aftermarket industry has been built around electronic program technology and the tweaking thereof.

HOLLEY PERFORMANCE PRODUCTS, INC.

Fig. 9.17: *A throttle body injection assembly. Throttle body units are used as original equipment by manufacturers to provide economical fuel injection capability. Aftermarket conversion kits are available as well.*

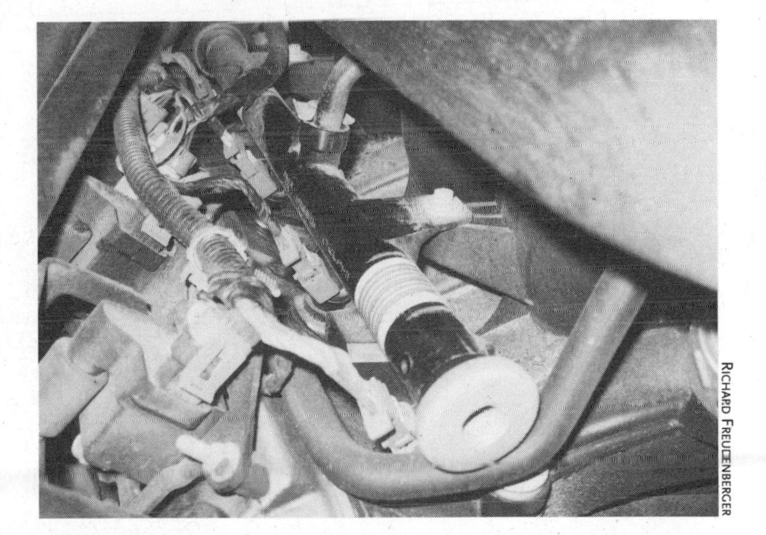

RICHARD FREUDENBERGER

Fig. 9.18: *Multi-port injection is more efficient than a TBI system, but is more costly because it requires more components. The fuel rail (dark tube, center) supplies the individual injectors, seen immediately to the left of the rail.*

Let's start with the injectors themselves. They're usually machined from high-strength steel and can handle alcohol fairly well, unless it contains excessive amounts of water. Carbon steel will oxidize, so the 170-proof ethanol you run through your Massey-Ferguson with impunity is not really appropriate for injected systems. Injectors are also susceptible to contamination from fine residual matter in fuel that would pass right through a gasoline filter and a carburetor. It pays to filter your fuel down to the micron level before introducing it to the tank of an injected vehicle.

In addition, injectors and their pumps are inevitably lubricated when using petroleum-based fuel. This isn't the case with alcohol, and although the lack of petroleum itself doesn't seem to cause a problem, water and acids, and other contaminants in home-brewed ethanol can take their toll over time. The addition of one bottle (6.4 ounces) of synthetic chainsaw or 2-stroke oil compatible with alcohol to ten gallons of fuel will inhibit corrosion and help lubricate the most sensitive parts for longevity.

Now we can look at some modifications. The mechanical approach is pretty straightforward, but is definitely not the crisp scheme one would hope for in a perfect conversion. Increasing the capacity of the injectors is a start. It's the equivalent of enlarging the orifice of a carburetor jet because it introduces a greater volume of fuel into the combustion chamber with each pulse. The capacity of an injector is measured in pounds of fuel per hour, so if you were seeking a 30 percent increase — a reasonable target assumption based on what we now know about alcohol fuel — you'd multiply the rating on your present injector by 1.3 (1 + 30% increase = 1.3). A typical four-cylinder engine might have an injector rated at 14 pounds per hour, at maximum flow. Adding 30 percent to that gives you 18.2, so you'll need to find a replacement injector in the 19 pound-per-hour range. The pulse control system will open the injector for the same period of time, but more fuel will pass

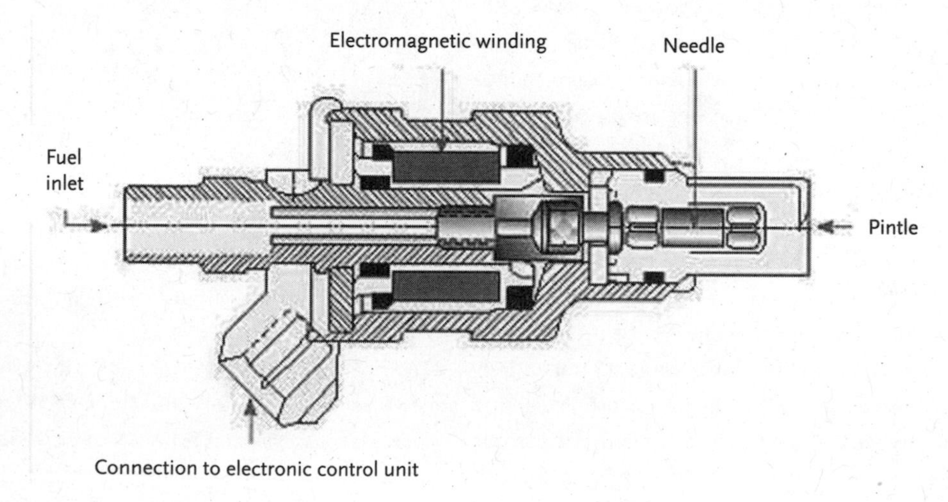

Fig. 9.19: *Cutaway illustration of a typical fuel injector. Injector tolerances are very close, so fuel must be thoroughly filtered.*

through the orifice. The oxygen sensor will take over and adjust the fuel flow based on the stoichiometric ratio as determined by the degree of oxygen in the exhaust stream.

Injectors can be modified in a machine shop, but in the long run it's cheaper and easier to simply buy replacements. As long as the injectors are accessible, actually switching them out is not difficult.

The fly in the ointment is that the system is still working with the stock fuel pressure, but you've enlarged the injector outlet. This alters the spray pattern, which affects atomization and lowers mileage. So, there's another, and better, way to get more fuel into the cylinders without bumping up the injectors, and that is to increase the delivery pressure. Fuel line pressure is controlled by a regulating valve placed in the common line downstream from the injectors. Holding the valve closed longer builds up pressure created by the injector pump. What's not used by the injectors is returned to the fuel tank through a return line.

Increasing the fuel pressure forces more fuel through the injector orifices during each pulse. It also has the effect of atomizing the fuel more effectively. Early Bosch Jetronic systems used in older fuel-injected Volkswagens and some domestic cars had adjustable pressure regulators. You can also buy adjustable regulators from aftermarket suppliers to replace the fixed-pressure regulator in the vehicle. Typical stock fuel pressures run from 28 to 30 some-odd pounds per square inch. Doubling that is well within the margin of safety designed into the system, since the fuel pump is built to produce three times as many psi. There is a pressure fitting in the common rail, and you'll need to buy or borrow a pressure gauge to get an accurate reading of where you're going with your adjustment. You shouldn't simply rely on an arbitrary pressure to achieve your setting; instead, you should use diagnostic tools like the exhaust gas analyzer mentioned earlier to seek the lowest CO reading, or establish the highest manifold vacuum with use of a vacuum gauge. Exhaust gas temperatures are also a very accurate measurement of complete combustion, as long as the readings remain within safe limits.

Heating the fuel as described earlier is also effective with fuel injection systems, since they don't have issues with vapor lock as carburetors do. Even warming alcohol to a level safely below its boiling point will help to thin it out and cause it to flow and atomize more effectively, increasing efficiency and fuel mileage. Temperatures between 140 and 160 are effective; there's no need to try and raise it any higher.

Now let's move on to electronic means as a way of modifying fuel delivery and combustion. In many ways, this is a lot simpler than fooling with plumbing parts because these are component kits developed for the high-performance aftermarket. Back in the day, street racers and competition gurus modified the electronic control units' circuitry to replicate what was available in the police packages sold to sheriffs' departments and highway patrol units across the country for their high-speed pursuit cars like the Camaro Z-28 and HP Mustang. Entrepreneurial types picked up on this and developed products that either replaced the MCUs or altered their configuration to achieve the desired goals: changing pulse width to adjust fuel mixture, advancing timing, or modifying airflow readings to trick the MCU into thinking more oxygen is available, forcing it to compensate by enriching the mixture.

These "tricking" or "fooling" devices are technically not legal for highway use, due to environmental regulations established for the automotive industry. So, they're simply marketed for "Off-Road Use Only" and clearly labeled as such. Universal kits are sold that work with most EFI engines. A control on the dash allows you to increase the fuel-air mixture at will once you've set it up, so it's a fairly simple way to utilize alcohol fuel and still be able to switch back to gasoline without a lot of bother.

There is also a second type of tricking device called a mass airflow tuner that alters the information coming to the MCU from the mass airflow sensor (MAF), a standard device in today's cars that reads the amount of oxygen coming into the engine, even compensating for temperature, humidity and altitude. The tuner intercepts this information and reprocesses it before sending it on to the control unit. Airflow tuners are more sophisticated than the simpler air-fuel enriching devices because they account for ignition timing and cover the entire range of engine conditions — idle, acceleration and full throttle — and can be adjusted at each level. It's important that the tuner you choose for alcohol fuel allows for a fuel increase of more than 25 percent. The tuners are series-wired in line between the airflow sensor and the MCU, and can be switched out of the system with the addition of a simple bypass circuit. The MCU will then default back to its normal readings.

An elementary form of airflow control was used on the second-generation Bosch Jetronic systems, which were electronic-mechanical hybrids used in certain Volkswagens and some Japanese imports. These do not require tricking devices because they can be adjusted at the airflow damper control right on the side of the unit. Advancing the contact arm re-establishes the relationship between the damper position and the pot resistor that sends its voltage signal to the electronic control unit, enriching the fuel mixture while the system still "thinks" it's running on gasoline.

The latest technology trend is toward fully programmable microprocessor control units that actually replace the vehicle's MCU. Again, these were developed for the competition market, but are ideal for alcohol-fuel conversions. A search of Internet sites or a quick cruise through the newsstand for high-performance magazine titles will turn up ample sources for these devices. Accel, AEM, and several other aftermarket manufacturers are the players in this game, and both domestic and imported cars are covered, for the most part.

Admittedly, this is a real enthusiast's approach, and just about every aspect of the engine's control and performance is open to modification. Internet user groups share information, and you can download data for your particular application and tweak it as needed, using a laptop computer and software provided by the manufacturer or online. Different data paths can be set up for different driving conditions or for alcohol fuel economy or high-performance modes, as well as for normal gasoline operation. Cost can be an issue, but, as with anything else, you get what you pay for.

Ignition Timing

Establishing the precise point at which the spark plug fires in relationship to the position of the piston is called ignition timing. Many factors, including the shape of the combustion chamber, the heat within the chamber, the speed of the engine, the compression ratio, and the octane rating of the fuel determine what timing setting is best. Timing can be set so the spark fires anywhere from 0° to 45° before piston top dead center (BTDC). The more degrees along the rotation of the crankshaft involved, the greater the timing advance is. As the degree setting moves back toward the zero or top dead center (TDC) point, ignition timing is said to be retarded.

There was a time when timing was established mechanically, directly from a gear on the engine crankshaft or camshaft. Engines with multiple cylinders use a distributor connected to this gear to direct high-voltage energy from the ignition coil to the individual spark plugs. At first, a driver-operated lever that was linked to the distributor controlled the timing changes, within the modest range necessary for engines of the day. Later, engine vacuums and centrifugal weights in the distributor more or less automated the process so no human input was needed. Much later, in the era when environmental controls became an important component of engine design, electronically controlled distributors were introduced that took input from engine-mounted sensors and made changes automatically and constantly as needed. Today's ignition systems have done away with mechanical distributors altogether and rely on the vehicle's microprocessor control unit to read, analyze and act upon information in a highly efficient real-time process.

But, back to the conventional distributor. A vehicle or engine-driven piece of equipment has specifications that let you know where to set static timing for normal gasoline operation. Static timing refers to the initial setting, at idle speed, before any dynamic timing advance from vacuum or weights comes into play. The distributor housing can be rotated by loosening a clamp at its base in order to advance or retard the timing of the spark in relation to the position of the crankshaft, and by default, the pistons within the cylinders. You'll need a timing light to read the degree marks at the crankshaft's harmonic balancer; the light's strobe flash is

RICHARD FREUDENBERGER

Fig. 9.20: *In the early days of motoring, ignition timing was controlled manually by a lever, shown here on this 1930 Model A Ford.*

synchronized with the firing of spark plug No. 1 and illuminates the degree marks quite clearly.

Beginning at the factory setting, first remove the hose to the vacuum advance unit and plug the hose with a plug or small dowel section. Turn the distributor to advance the reading at engine idle to about 12° or 15°. Lock down the distributor at this point and reattach the vacuum hose, then drive or operate the engine under load, and at high speeds. If the distributor advance is too great, the engine will ping loudly on acceleration and misfire at higher rpms. Remember, the static setting is, say, 12°, but the dynamic advance built into the distributor can add an additional 15 to 25 degrees. The point is to run the engine with the highest degree of advance allowable without deteriorating performance or damaging the engine with pre-detonation. If audible sounds of pinging or knocking occur, you'll have to retard the timing by a degree or two and re-test. Continue this procedure until the pinging ceases. With some distributors, there isn't enough range built into the housing to allow maximum advance. In such cases, you can lift the distributor housing from the engine block after removing the hold-down clamp and bolt, and rotate the distributor shaft one tooth forward (in the direction of rotation) to gain the advance needed. Then go through the testing procedures and lock down the distributor when finished.

This is a decidedly unscientific method of setting timing, but it functions well in lieu of a knock sensor, an auditory device that is built into modern electronic systems and is available as an aftermarket purchase. Alcohol can burn in a wider range of air-fuel mixtures than gasoline can, and has a higher octane rating, which allows it to operate safely with much higher compression ratios. Running at the highest allowable advance lets the engine run cooler, attain better fuel mileage, and develop more power under load. High compression, proof strength of the alcohol fuel, and high-energy ignition (HEI) systems all affect the degree to which advance can be dialed in.

Electronic ignition systems have the disadvantage of not being as accessible, or modifiable from a backyard mechanic's point of view, as the old mechanical distributors. However, with a bit of investment, there are several aftermarket options open to the tinkerer that will co-exist splendidly with alcohol fuel setups. Some electronic ignition kits made for retrofit onto the earliest engines have a wide range of advance, enough to accommodate gasoline and alcohol fuels. There are also electronic devices made for the performance market that allow either manual control of spark timing or use of a programmable path that lets you set the system up for whatever your situation might be with regard to proof strength, engine load, compression ratio, or fuel mixture.

Spark Plugs

Changing or readjusting spark plugs for alcohol fuel may offer a slight benefit in some cases, though the advantages do not seem to carry across the board to all engines. In some of our early testing on small single-cylinder engines, reducing the spark plug gap by .002-inch seemed to aid in starting, but that did not hold true with the larger multiple-cylinder engines. After working with a variety of small air-cooled and motorcycle engines, two-stroke models, automotive, industrial and aircraft engines, the conclusion was reached that alcohol-fueled engines responded better to changes in heat range rather than to any adjustment in electrode gap.

Spark plug heat ranges may mean nothing to the average motorist, but to mechanics, motorcycle enthusiasts, and competition drivers, finding the right heat range is vital. The plug's configuration and the depth of its insulator determines how much heat is retained in — or carried away from — the cylinder head while the engine is at operating temperature, and the type of fuel, and how it behaves in the environment of the combustion chamber, can have a noticeable effect on how well the plugs perform in that environment.

In a healthy engine, the heat range is chosen to fully combust the fuel that's available under the operating conditions the engine is subjected to. Frequent stop-and-go driving demands a hotter plug to burn away deposits that would otherwise be cleansed with long runs at highway speeds. The same spark plug would probably be too hot for high-speed driving because the engine develops higher temperatures at speed. In an older engine that may be combusting some oil along with the fuel, a hotter plug will help to burn away oil deposits and prevent fouling.

Ethanol's cooler burning characteristics would seem to favor a higher heat range than gasoline, but if you choose to experiment, don't simply make assumptions. Check the appearance of the spark plug frequently to make sure it isn't burning too hot — ceramic cracking and a glassy sheen are telltale signs. Cool plugs will simply foul easily, but a spark plug in a heat range too high for conditions can do serious damage to the engine's piston and cylinder head if left uncorrected.

RICHARD FREUDENBERGER

Fig. 9.21: *The appearance of the spark plug electrodes — specifically, the bent ground electrode and the small center electrode tip beneath it — can be used to diagnose conditions in the combustion chamber. This example shows a slightly rich fuel mixture and a small amount of oil fouling.*

Hot Air Induction

Heat induction, or preheating of air before it enters the intake manifold inlet (or carburetor on older vehicles) helps to vaporize the fuel when the engine is cold. Heat is created in the exhaust manifold from the moment the engine is started, and a sheet metal duct system directs it to the manifold inlet. A thermostatically controlled damper shuts off this source of warm air once the engine comes to temperature, so cool, denser air can be used for normal combustion.

This damper is usually operated by a bimetallic spring or by a vacuum controlled by a switch mounted elsewhere on the engine. You can either fasten the damper flap so it always directs warm air to the inlet, using a spring or sheet-metal screw, or you can locate a constant vacuum fitting on the engine and use that to keep the damper in the active position. Be sure to test the fitting at both idle and running speeds to make sure it's ported to operate constantly, not just at idle.

If you're working with an engine that does not have any heat-induction device, you can probably find one to fit at an auto salvage yard, or make one by fabricating a sheet metal shield to cover the top and side of the exhaust manifold or part of the exhaust pipe if necessary. Braze or weld a metal collar to the top of the shield sized to match the collar already beneath the inlet horn on the vehicle's air cleaner housing. If there is no collar, you can cut a hole in the housing and add one; pre-made exhaust and tubing fittings in various diameters are usually available at auto supply houses. A diameter of 2 inches or 2¼ inches should be sufficient for cars and trucks. Buy a length of flexible exhaust tubing and fasten it between the fitting on the manifold shield and the air cleaner using a clamp or sheet metal screws.

Fig. 9.22: *An air preheating system fabricated for the ethanol carburetor of the 6-cylinder Chevrolet van. Inlet temperatures drawn from the engine's cooling system hover in the 175°F range.*

BackHome photo collection

When making your own preheating plenum, be sure to introduce the warm air to the inlet housing before the filter element, and choose the diameter of the pipe to match the size of the snorkel already on the housing to avoid starving the engine of air. On vehicles with electronic feedback carburetors or fuel injection systems, the air must be introduced downstream of the air temperature sensor so it doesn't pick up an artificially false reading.

Preheated air contains less oxygen. But since ethanol already contains a significant amount of oxygen, the effect of warm air not only atomizes the fuel more efficiently, it increases fuel mileage as well. A preheating plenum adapted from a Ford Pinto heater core housing and used in our six-cylinder Chevy van took heat from the cooling system at 175°F. Exhaust manifold preheaters can achieve temperatures much higher for even better results, but temperatures should be limited to around 350°F or preignition may occur.

Cold-Starting Systems

One point consistently raised by alcohol-fuel opponents is the fact that ethanol does not vaporize as easily as gasoline, and needs some assistance in cold temperature start-ups. E-85 blends have solved the problem by mixing the ethanol with 15 percent gasoline, which is enough to set off the fuel charge in most winter climates. Those in particularly cold regions get a special winter blend that contains up to 15 percent more gasoline to lower the flashpoint even further.

Still, the issue remains that for people using straight alcohol fuel, temperatures below 40°F or so can be a problem. Carbureted engines have the most difficulty because the fuel delivery is not pressurized. Fuel-injected engines, because of their fuel spray, will start at lower temperatures.

The simplest and least expensive way to provide the flash needed for cold starting is to inject a small amount of gasoline into the manifold at startup from a remote reservoir. Some of our earliest conversion projects used a 5-gallon tank and an electric fuel pump to send a spurt of gasoline through a fuel line and metering jet fastened to the air cleaner housing. Cautionary letters and howls of protest from safety advocates pointed out that this was too risky in the event of a backfire through the carburetor (which isn't all that uncommon), so the newer modifications inject the fuel further downstream, into the manifold or even more conveniently, into the positive crankcase ventilation (PCV) hose which leads directly to the manifold and draws a vacuum while the engine is running.

Plumbing this kind of setup is relatively easy. To start, you really only need a small canister, about the size of a windshield washer reservoir or lawn mower tank. Mount the tank under the hood, away from the exhaust manifold or other source of direct heat, and route a ¼-inch fuel line to an electric fuel pump bolted to the firewall or other convenient location. Connect another section of fuel line from the pump outlet to a tee fitting in the PCV hose. The tee can simply be 3/8-inch brass with two 3/8-inch hose barb fittings threaded into the holes opposite each other. The branch can be fitted with a third hose barb (or a fuel line fitting if you're using steel line) that's been modified by soldering a small-orifice metering jet onto its end. Make sure the jet's head is smaller than the diameter of the threads and is bonded to the fitting securely, for you do not want it sucked into the manifold if it comes loose. Don't be afraid to drop down to a ¼-inch thread for the branch if needed to utilize a smaller jet. Tee fittings are made with many types of thread combinations.

BACKHOME PHOTO COLLECTION

Fig. 9.23: *The positive crankcase ventilation system (a portion of which is shown here, where the hose and valve is seated in the valve cover) provides a safe entry point for a spurt of gasoline used in cold starting.*

You don't even need to use an electric pump for the small amount of fuel we're talking about. I have seen foot-operated windshield washer systems converted to gasoline service (make sure the plastic components are compatible with gasoline; many are), and pumps adapted from camp stoves that use white gas or lantern fuel are ready-made for this purpose. Make certain that whatever type of pump you use is self-closing so that fuel is not sucked into the manifold while the engine is running.

If you're concerned about safety, tractor and auto supply houses sell commercial diethyl ether injection kits for installation on diesel tractors and trucks. These are very common in cold climates, and the kits include the ether canister, an electronic solenoid valve, an atomizer, a momentary switch, and fuel delivery line. Some even use a thermostatic sensor to open the circuit if the engine block is warm, shutting off current to the valve when ether is not needed. These types are usually wired into the starting circuit of the vehicle's ignition switch. The ether line is plumbed into a hole tapped into the intake manifold, and a measured amount of starting fuel is injected at startup.

Other creative cold-starting methods have used propane torch canisters and solenoid shutoff valves from LP-powered equipment such as generators and forklifts, and glow plugs from diesel tractors to heat alcohol fuel injected at the manifold. Preheating the alcohol in cold weather can be effective, as well.

Compression Ratio Boost

On a somewhat higher tier, expense-wise, in the modification and conversion process is raising the engine's compression ratio. Because of alcohol's high octane rating (around 106, or 19 points higher than unleaded regular) it can safely cope with compression ratios up to 16:1 without symptoms of preignition. High compression ratios make very efficient use of the fuel by increasing the velocity of the flame front on ignition and developing greater cylinder pressure, which translate to increased power.

The reason all manufacturers don't utilize higher compression in standard engines is that low-octane gasoline precludes it. Current passenger-car ratios are in the range of 9:1, which is an acceptable compromise between the grade of fuel available and high-efficiency combustion. A boost in compression can raise horsepower by 10 to 30 percent or more, and increase mileage by up to 20 percent.

The simplest way to raise engine compression is to mill the cylinder head, a procedure that involves machining a small amount of material from the head's surface, where it meets the engine block. This is done routinely any time the head is removed to assure that it's perfectly flat, albeit not usually to the degree that we're talking about for a compression boost. Planing the engine block in the same manner is a more radical step that, together with the milling, may yield an increase of up to two compression points, which is adequate but not optimal.

Milling the block and head surfaces maintains the stroke, or the distance the piston travels, at its original specifications, so no major rebuilding is involved. The same holds true for installing a set of high-compression pistons, which are shaped to pack the air-fuel mixture tightly into the combustion chamber and improve turbulent flow. A competition speed shop or auto machine shop should have technical specs on high-performance pistons, and can tell you if they're available for or adaptable to your particular engine.

As the compression ratio increases, generally above a point or two from the standard, issues arise as to the integrity of the connecting rods and crankshaft, and particularly the bearings connected to them. At competition-level ratios of 13:1 and higher, main bearing journals, head bolts, and even the pattern and number of bolts used in the design become critical issues. Again, a good machine shop will have knowledge on these points, but a guide to follow is that generally, domestic engines manufactured prior to the late 1970s, particularly V-8 engines, are designed with stout bottom ends and robust components, and will accept a few points of compression ratio increase without whole-hog modification. Certain stone-solid, long-tenure sixes, such as Mopar's slant-six or Ford's 4.9-Liter straight six, were undeniably overbuilt to begin with and can take a lot of punishment without failing. As smaller-displacement, higher-revving engines were developed, the "overbuilt" approach was eschewed in favor of cost cutting and utilizing improvements in metallurgy.

For someone not regularly occupied in rebuilding engines, going the route of a full custom rebuild is very costly and probably not worth the expense in the time and labor involved unless money is not a major issue. In high-compression engines, starters, batteries and charging systems also may have to be beefed up to handle the added pressure. In some cases, using a good factory short block or complete working engine from the "muscle car" era of the late 1960s and early 1970s may be feasible, depending upon its compatibility with the vehicle it's to be used in. A small-block high-output engine can still be an economical alcohol-fueler in a ½-ton pickup or mid-size Detroit sedan. These domestic hot rods, and some performance-oriented imported cars of the 1980s genre designed to burn premium fuel, can make excellent platforms for building alcohol-fueled engines without going to the expense of completely re-designing internal components, assuming the engines are in first-rate condition to begin with.

Diesel Engine Block Conversions

Before moving on to turbocharging as a means of boosting compression, I want to address the use of converted diesel engines for alcohol fuel use. Rather than trying to build a high-compression block from a stock gasoline engine, this approach takes an already substantial factory-built diesel engine and converts it for spark-ignition use. It's not a low-cost conversion by any means, but it immediately resolves the litany of issues to be faced in modifying a gasoline engine for super-high compression.

Essentially, the diesel block is fitted with gasoline-service components. Some engine blocks are cast with an eye toward sharing as many parts between gasoline and diesel as possible, mainly with regard to casting bosses and bolt patterns. If this is the case, the task is easier, but beware of the GM automotive diesels of the late 1970s, which were modified from gasoline blocks and are somewhat problematic in their durability.

In the US and elsewhere, research has been done in this area. Typically, the engine is de-compressed from a ratio of 21:1 down to a level more compatible with alcohol; spark plugs are fitted into the glow plug ports; and an electronic ignition system is installed. Stock diesel injectors cannot tolerate ethanol without lubricant, so multi-port injection at the intake manifold provides the fuel. Other modifications, mainly to make the sensor readings compatible with the electronic control module, are fairly straightforward. Realistically, this is beyond the ken and probably the budget of the average shade-tree mechanic, but it's encouraging to know that the possibility exists.

Super Aspiration

There is another way to increase the compression ratio in an engine, through the use of a supercharger. Supercharging is the term used to describe the introduction of compressed air into the intake manifold. The benefit of this "super aspiration" is not only an increase in engine compression, but the fact that the compressor delivers warm, high-velocity air to help atomize stubborn fuel particles that otherwise might not disintegrate effectively.

Technically, all such compressors are superchargers, but they're further broken down into classifications based on how they operate. Turbochargers are driven by a miniature impeller powered by the engine's exhaust gases. They are effective on higher-speed acceleration and at wider throttle openings. Traditional superchargers are usually belt-driven off the crankshaft, though some are geared. They are either positive displacement pumps, as in the Roots or twin-screw types, or centrifugal

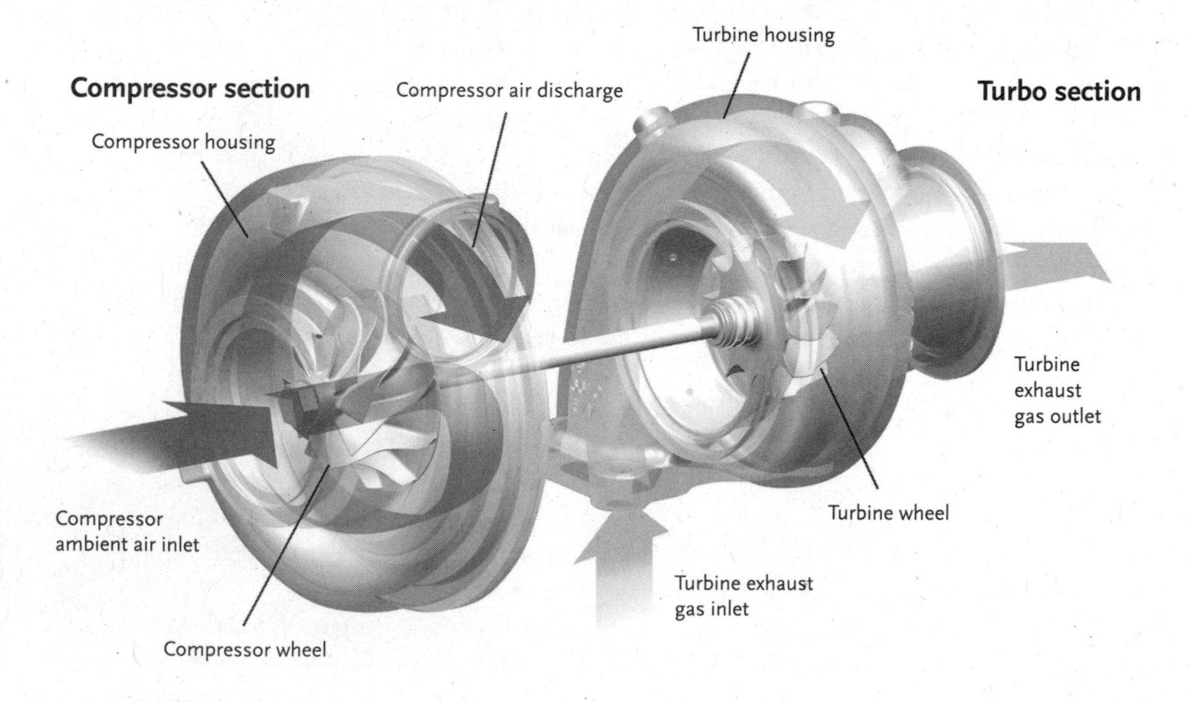

Fig. 9.24: *A turbocharger uses the force of the engine's exhaust gases to drive a turbine wheel, which is joined to a compressor that forces air into the engine manifold under pressure.*

compressors similar to the compressor turbine used in the turbocharger. The positive displacement blowers give a substantial boost in the low and mid-range speeds, but taper off at the high end. The centrifugal types behave much like turbochargers and function better at high rpms.

The great feature of forced aspiration is that the engine's mechanical compression ratio can remain at normal levels to accommodate starting and idling. Since all the super aspirators are adjustable to some degree — in turbochargers by controlling the wastegate that governs how much exhaust drives the compressor, and in superchargers with a bypass valve — the boost doesn't have to kick in until the engine has developed some rpms. Boost pressure can be governed by a knock sensor that protects against over-boosting, but in any event, pressures suitable for gasoline can be increased substantially when running on alcohol fuel. High boost pressures, like high compression ratios, still call for beefed-up engine components. Like the more involved mechanical compression modifications, supercharging does not come cheaply, but the installation is not nearly as difficult, especially with the availability of aftermarket kits.

Small and Single-Cylinder Engines

While I've mostly been discussing automotive and multi-cylinder engines, it's appropriate that we consider smaller engines as well, since they power the mowers, pumps, chainsaws, garden shredders, snow throwers and portable generators we've come to rely on so heavily. The principles are essentially the same, but there are a few distinctions worth mentioning.

Four-Stroke Engines

For the most part, single-cylinder four-stroke engines use a simple sidedraft carburetor to get atomized fuel into the manifold. With many engines, the carburetor and manifold are cast as one piece, simplifying the design even further. These types of engines generally run at a fixed speed, so issues of off-throttle performance and accelerator circuits are negligible. All we really care about is starting, idling and efficient running.

Adjustments involve only two screws, at the main jet and at the idle jet. With a few exceptions, both these jets are fully adjustable. The main jet screw (usually the one with the largest head) should be opened one full turn to begin with, and the idle jet opened by ½ turn. It's not necessary to change the setting on the idle speed screw, which is threaded through the throttle lever in the center of the carburetor.

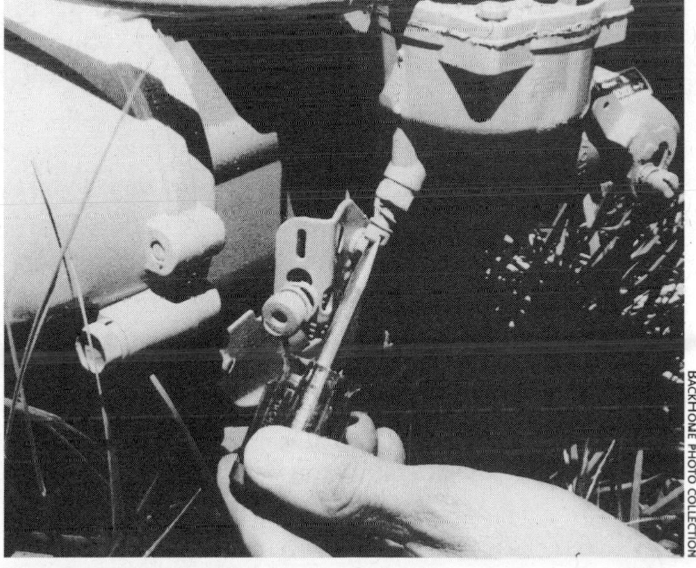

BACKHOME PHOTO COLLECTION

Fig. 9.25: *Adjusting the main jet on a small four-stroke portable generator engine.*

Once the engine is started (you may have to squirt a dollop of gasoline into the spark plug opening or the carburetor inlet to do so), you can fine-tune your adjustments. First bring the engine to normal operating speed and slowly draw up the main jet screw by turning it clockwise, one-quarter turn at a time. When the engine begins to miss or hesitate, back off until it runs smoothly again, then continue back a fraction more. Drop the speed down to idle and back out the idle circuit screw. If the engine runs roughly, tighten the screw slowly until the idle smooths out; if it's drawn down too much, the engine will stall. If you can't get the idle circuit setting right, you may have to increase the idle speed slightly, but do this only as a last resort.

Like the larger engines, these small jobs will benefit from a few simple modifications. Ignition timing is usually fixed on the economy engines, but if not, there may be some adjustment in the timing plate behind the flywheel. You'll need a wheel puller to remove the flywheel after you've unthreaded the flywheel nut (do not hammer on the crankshaft!); the plate is usually secured by two screws. Rotate it against the direction of crankshaft travel to advance timing slightly, then reinstall the flywheel. If the engine balks or offers resistance at starting, you've probably advanced the timing too far and will have to back it off.

The most effective modification you can perform is to heat the inlet air or manifold of the engine. Heating the fuel easily causes vapor lock, probably because of the small fuel-bowl design. To heat the inlet air, you can shroud an inlet pipe the same diameter as the collar on the air filter mount around the muffler with sheet metal. Then re-mount the cleaner at the inlet end of the pipe and fasten the outlet end to the carburetor, where the cleaner originally was. A better system, for engines with separate manifolds, is to fabricate a sealed metal housing for the intake manifold and route exhaust gases through it before piping them to the muffler. This will assure that the incoming fuel-air charge is nicely vaporized at elevated temperatures, without bringing about vapor lock.

It may not be worth it to even bother with a cold-start system, since a can of ether is such a convenient option. However, applying the simpler starting-aid techniques from the cold-starting section above will work just as well on small engines as it does on larger ones.

Two-Stroke Engines

Two-strokers are different animals than their four-stroke cousins, but the carburetors function in much the same way. Due to strict weight requirements in most two-stroke applications, the carburetors are bolted directly to the inlet port of the engine. They have high-range and low-range adjusting screws, corresponding to the idle and main circuits in the four-stroke engines. If you've ever used a chainsaw or weed cutter, you may have noticed that, just as the engine is about to run out of fuel, it will speed up before conking out completely. This is what happens with the fuel metering adjustments — an over-rich setting will cause the engine to smoke, stumble and run at reduced rpms. Leaning out the mixture eliminates those symptoms, but if it's leaned too far, the engine speeds up to an unsafe point. When adjusting the high range, start rich and work toward the goal of smooth running. If you go beyond that point and rpms increase, back the screw off to a safe level. The low range can be set in the same manner, with the target being a happy medium between a rich mixture and one that's too lean.

Petroleum-based two-cycle oils do not mix or burn well with ethanol fuel. What's worse, the lower the proof strength, the more the problem is exacerbated. That's a shame, because all two-stroke engines require that the lubricant be mixed with the fuel, otherwise the engine will seize up. The next section will address synthetic lubricants in particular, but here I'll mention that the additional cost of synthetic two-stroke oils are both worthwhile and necessary. A lot of synthetic two-stroke products have been developed for competition motorcycles (to run on ethanol) and for motocross dirt bikes. These have been around for a while, sold under brand names such as Amsoil, Klotz, and Red Line. Castor bean oil is also an old favorite and plant-based as well, and some more recent vegetable-based lube mixes have come onto the market too.

Because motorcycle racing has a history of alcohol fuels, oil manufacturers should offer some guidance as to the proper mix ratios for methanol and ethanol. As a rule, the mix ratio is more diluted (i.e., more fuel in the mix) than it is for gasoline, and the spark plugs and combustion chamber will remain clean for a longer period of time.

Using Synthetic Oils

As mentioned above, synthetic lubricants are a better choice to use with alcohol fuels than standard petroleum-based oils. While it's true that synthetic oils can cost two or three times more than conventional oil, there are a few good reasons why synthetics are preferred in your crankcase.

For one, synthetic lubricants have an unusually high film strength, due to their long-chain chemical structure. In practice, this means that the lubricant film between metal surfaces such as bearings and crankshaft journals is particularly tenacious. Also, the viscosity range in synthetic oils is fairly broad, meaning that the oil remains thin at cold startup, the more easily circulated through galleries and bearing parts, and thickens as the temperature rises, providing protection at operating speeds. Furthermore, synthetic lubricants are formulated to not degrade easily at high temperatures, extending their lifespan considerably. It's routine for synthetic oil to be left in the crankcase and only the filters changed for periodic maintenance until the oil has run through a half-dozen or more maintenance cycles.

Factory Flexible Fuel Vehicles

Flexible Fuel Vehicles, or FFVs, have been manufactured in the US since 1994. At this writing, there are more than 6 million FFVs on the road, and they include sedans, SUVs, and compact and full-size pickups from every domestic manufacturer and Nissan, Toyota, Mazda, Isuzu and Mercedes-Benz.

Flexible Fuel vehicles like my Ford Ranger are priced at the same level as comparable gasoline-only cars and trucks, but can use unleaded regular, the 85 percent ethanol/15 percent gasoline fuel sold as E-85, and any

RICHARD FREUDENBERGER

Fig. 9.26: *The Flex Fuel Vehicle emblem shows a stylized highway and corn stalk. Currently eight automobile manufacturers offer FFV cars or trucks.*

combination of the two without doing anything more than turning the ignition key. The only indication that my truck is any different from any other Ranger is a tiny green cornstalk emblem pasted to the lower corner of my tailgate, yet when the truck was new in 2000, a buyer was eligible for generous federal tax credits, and in many places, state credits as well.

From a technical standpoint, FFVs use the same technology as other cars. Initially, the main difference between a Flex Fuel vehicle and a conventional gasoline vehicle was the addition of a fuel sensor that detected the ethanol/gasoline ratio and made adjustments for it. Other components such as the fuel tank, fuel lines, and fuel injectors were modified slightly. Later technology allowed manufacturers to drop the fuel sensor and simply use the existing oxygen sensors to deliver the needed data. The computer chip in the electronic control module/powertrain control module (ECM/ PCM) is reprogrammed to broaden its control range, which covers ignition timing, injector timing, and other pertinent factors. The fuel pulse width is increased on alcohol and returned to normal range on gasoline, and higher fuel pressure range is indicated, at least on the Ford products.

It has only been since the State of the Union address in 2006 (and the associated support for ethanol fuels) that manufacturers have actively marketed their FFV lines. Yet the Flex Fuel vehicles make excellent candidates for 100 percent alcohol fuel conversion because there is very little that has to be done to them. The main issue is cold-starting on "neat" or 100-percent alcohol, and I've addressed that earlier. Other modifications favored by ethanol enthusiasts are fuel heating to increase viscosity, or bumping up the fuel line pressure.

One challenge remains, and that is that the oxygen sensor will activate the "check engine" light on pure alcohol fuel, because the mixture is burning leaner than the stoichiometrically correct programming allows. So, another piece of valuable trick technology is an O_2 tricking device, a fairly inexpensive item that modifies the signal to the electronic control module and allows the engine to run at a leaner air-fuel mixture than otherwise allowed. This not only increases alcohol-fuel mileage, but also virtually eliminates hydrocarbon emissions without drastically bumping up undesirable oxides of nitrogen. It also keeps the "check engine" light off.

Drivers of conventional ECU-modulated vehicles have been lured to plug-and-play Flex Fuel conversions by aftermarket add-on technology made popular in Brazil in the last decade. The control boxes, marketed in the US under names such as FlexTek, Full Flex, and

Fiat's TetraFuel System

For those skeptical about the ability of manufacturers to develop practical, economical Flex Fuel vehicles, one only need look at the Fiat Siena TetraFuel developed as a joint venture between Fiat Group and a subsidiary, Magneti Marelli of Hortolandia, Brazil. The 1.4-Liter four-cylinder Siena has been modified by Marelli to burn ethanol, gasoline, compressed natural gas, and any mixture of gasoline and ethanol, and it will switch between the fuels automatically, adjusting ignition timing and fuel pulse width as needed without sacrificing emissions control.

This TetraFuel version of the Siena is sold only in Brazil, where the cost of fuel is a big concern and motorists routinely shop for vehicles based on what their fuel costs will be. Of course, Brazilian society is now accustomed to the availability of diverse fuels, so the four-fuel option fits their needs to a "T". Compressed natural gas is the least expensive auto fuel in the country, and it is fine for cruising at speed. When conditions call for more power, the liquid fuel will kick in to provide a boost for passing or climbing a grade. The liquid fuels are held in a regular 12.6-gallon tank, and the compressed gas in two cylinders in the trunk. The engine has a set of injectors for the liquids and another set for natural gas. The TetraFuel system is designed to reduce fuel costs by up to 40 percent when compared to a gasoline-powered car, but it reduces greenhouse gases by up to 24 percent as well.

1.4 TETRAFUEL

FIAT AUTOMOVEIS

FIAT AUTOMOVEIS

Fig. 9.27: *Fiat's TetraFuel system allows the new Siena to operate on gasoline, ethanol or compressed natural gas.*

White Lightning, are specific to four-, six- and eight-cylinder vehicles. The units are wired in line between the electronic injectors and their individual cables leading to the ECU, using the existing male-female receptacles. They take information from the ECU, reprocess it to suit the ethanol data stored in their memory, and send the modified signal to the injector, which adjusts the pulse width accordingly. Switching the box off essentially disconnects it from the system, allowing the normal signals to travel between the ECU and the injectors for gasoline operation. There's no question that these trick Flex boxes work, but reading a few posts on the Internet may raise questions as to how well. The biggest complaints seem to be that the units have difficulty discerning fuel blends, and that they function best when using a major percentage of one fuel or the other.

Alcohol and Diesel Engines

Though spark-ignition engines have been the crux of alcohol-fuel research, compression-ignition diesel engines have not escaped scrutiny, primarily because they are highly efficient and very durable. I talked about diesels briefly, a few sections earlier, in pointing out that they have extremely high compression ratios, which is essentially what heats, and ignites, the fuel.

There are several immediate problems with using ethanol as a fuel in diesel engines. First, the anti-knock qualities that are so attractive in alcohol are actually a drawback when compression is igniting the fuel charge. Diesel fuel has a cetane rating, which is used to gauge how easily it knocks, precisely the opposite of the octane rating. Alcohol ignites under compression at temperatures approximately 60 percent greater than diesel fuel does. Alcohol also does not have the lubricity needed for the injectors and injector pumps in a diesel engine. As anyone who's handled diesel fuel knows, if anything, it's oily. And, the diesel injector pumps do not tolerate water.

Still, alcohol has been used proportionally in diesel-fueled engines, and research continues in this field, particularly for the fact that diesel emissions can be significantly reduced with the addition of alcohol in combustion. Back in the fuel-crisis days of the late 1970s and early 1980s, so-called fumigation systems were popular in agricultural circles. These aftermarket adaptations used carburetors mounted on the engine's intake manifold. After adjusting the injection system to cut back fuel delivery and tying in the carburetor's throttle valve to match the tractor's load, the engine essentially took between 30 and 50 percent of its fuel charge from the carburetor, depending on load. At idle, the engine consumed diesel almost exclusively.

For turbocharged diesels, an alcohol injection system was developed at around the same time by an outfit called M & W Gear Company. This used an alcohol injector mounted downstream from the turbocharger. It discerned turbocharger pressure and delivered the appropriate amount of fuel to lower diesel consumption and boost horsepower.

Some years ago, in São Paulo, Brazil, Mercedes-Benz engineers were eager to show me their ethanol-fueled diesel buses, which they'd developed as part of the state alcohol program. Using a cetane booster made from

Fig. 9.28: The E-95 BSR Saab 9-3 utilizes ethanol in a diesel engine to reduce emissions by approximately 95 percent.

cane-based alcohol — triethylene glycol dinitrate — they blended a small percentage with straight ethanol and used the fuel in city buses. To address the lubrication issue, the injection pumps were fitted with a separate filtered oil line from the oil pump and a return line to the crankcase. A small amount of oil actually gets into the fuel to help lubricate the injectors, but not enough to impact the reservoir in the crankcase.

Early in 2008, the Swedish automotive-tech company BSR Svenska AB unveiled their ethanol-powered diesel conversion, a Saab 9 3 optimized for E-95 (95 percent ethanol, 5 percent gasoline) fuel. The engine's combustion chamber was modified and the fuel system and ECU software altered to suit the properties of alcohol. The 195-horsepower engine achieved fuel mileage figures of around 46 miles per gallon and reduced carbon dioxide and hydrocarbon emissions by approximately 95 percent. This automobile is not in production, but has provided a clear path for future development. The company markets E-85 conversion kits for a variety of gasoline cars.

Other experimenters have blended ethanol and plant oils to make diesel fuel with varying success. Mixing hydrated alcohol with castor oil was not an uncommon practice among fuel alchemists some 30 years ago. Blended in a 1:4 ratio with ethanol dominant, the fuel was passably acceptable, though difficult to keep consistent. There is no documentation that I'm aware of that indicates its effect on engine or component longevity. More recently, blenders have combined biodiesel and anhydrous alcohol, which allows a fairly broad range of proportion. This also addresses the lubricity issue of using ethanol, and the problems associated with water in hydrated alcohol. The prudent among us will approach these experiments with eyes wide open, because of the potential damage that unsuitable fuels can wreak on costly diesel engines.

Space Heating Systems

Kerosene, or fuel oil, is still a major component of the heating industry and more common in some parts of the country than natural gas or propane. Heating fuel has the same characteristics as diesel fuel, so it burns with a distinct petroleum odor and leaves an oily residue when it's done.

But fuel oil burners can easily be adjusted to burn alcohol, and with considerably less effort than it takes to modify a vehicle engine. The burners in oil furnaces are called gun burners because they're simply barrels fitted with a nozzle that injects pressurized fuel into a refractory-lined combustion crucible. An electric spark ignites the fuel spray, and the resultant heat is directed through tubes in a heat exchanger. Combusted gases go up the flue pipe and heated air is forced into the duct plenum.

The burner conversion process is very simple. Because alcohol already contains oxygen, you have to reduce the amount admitted to the burner by turning the air control down to limit draft. It may be a slotted collar or a rotating plate, but the point is to cut down the air supply to replicate the normal flame pattern. Then, because ethanol has a lower heating value in Btu's than fuel oil, you have to increase the size of the metering nozzle to get an equivalent amount of heat output.

Unthread the nozzle with a box wrench and take it out. You won't have to drill this orifice, because burner nozzles are available in several sizes, based on gallon-per-hour flow rates. Get one with a diameter about 15 percent larger, or the equivalent of a 35 percent increase in rate. Some conversions are more effective when the nozzle's spray angle is changed as well.

Though each manufacturer's wares are different, one constant is that the oil pump will probably not get the lubrication it needs from alcohol fuel alone. For this reason, it's necessary to add up to 5 percent kerosene or fuel oil to the new fuel to provide lubricity for the pump. Advocates of renewables may substitute biodiesel for the petroleum-based fuel.

TEN

Case Studies

Farm Ethanol: Painterland Farms, Westfield, Pennsylvania
• Cellulosic Ethanol: Iogen Corporation, Ottawa, Ontario

On-Farm Fuel Production

Tucked almost against the New York border in one of Pennsylvania's northernmost counties, Painterland Farms is an organic dairy operation that's home to some 300 Holsteins, Jerseys and Linebacks. Several generations of family work the farm, with help from hired staff.

In 2005, John and Lynda Painter and son John, Jr. began work with their Cooperative Extension Agent J. Craig Williams to develop a model for a $15,000 USDA Sustainable Agriculture Research and Education (SARE) grant that would provide on-farm research to establish an enhanced dairy cattle diet and potential fuel sustainability on a working dairy farm. The primary intent was to explore the practicality of implementing a small-scale ethanol facility within a stable operation to

PAINTERLAND FARMS

Fig. 10.1: *John Painter, Sr. standing by the distillation column of Painterland Farms' alcohol fuel plant.*

broaden the scope of value-added products and to supply contemporary energy needs through traditional distillation techniques and the introduction of new disciplines.

The Painters maintained that even in today's complex markets, an established, diversified operation would be able to provide for many of its own needs, working toward sustainability while protecting its ability to maintain operations during periods of market volatility and economic upset. The project took on two main goals: (1) to prove the viability of an enhanced dairy cattle diet through the feeding of distiller's grains provided and improved by the distillation process; and (2) to provide a clean, renewable, on-site energy source in the form of ethanol. A third objective was to minimize waste by using salvaged components where possible, and recycling water and heat within the system to economize on the input of resources.

Since maintaining a successful farm operation was fundamental to the project, the participants had to estimate the total volume of distiller's grain needed annually. Then, based upon that figure, they were able to establish a design volume needed for the distillation equipment.

After some searching, a used distillation column and equipment were located in North Carolina and brought back to Pennsylvania, where it was set up in an unused dairy barn. The Painters modified the barn by raising a portion of the roof to accommodate the distillation tower, and they cleared out the inside to make room for mixing equipment, fermenting vats, a beer tank, and distiller's grains processing.

The Painterland ethanol operation uses organic corn and barley, and, at some point in the future, it will use rye as feedstock. The market price of these commodities determines, to a great extent, which crop will be used. If the price of one rises to the point where it's more valuable on the market than it is as a fuel, another feedstock will replace it for distillation. This kind of flexibility helps to keep the price of their ethanol down, and also keeps the Painters on their toes scrutinizing commodity markets and planning for future crops. Similarly, the fuel source for the steam boiler is currently rice coal, a fine-grain anthracite common in the Pennsylvania countryside. It's readily available, but as a fossil fuel is subject to market fluctuations. The family is investigating renewable resources and biomass options to supplement, or eventually replace, the coal source that could become problematic in the future.

From Feedstock to Fuel Stock

The process begins with ground corn and grain in bin storage. Shredding, grinding or mashing gives the fine meal grind needed to initiate the breakdown of starches. A conveyor chute transfers the material to a ration mixer in the distillery, which includes a set of scales to record weight measurements.

The grain is then piped to a group of stainless steel vats, where cooking and fermenting processes occur in the same vessel. Using a relatively new raw starch hydrolysis or "cold-cooking" method, the Painters are able to complete the enzymatic breakdown of starch into sugar at near room temperatures — about 90°F — which eliminates two stages of the traditional process and saves considerable energy. Genencor International Inc. has been marketing their Stargen™ granular starch-hydrolyzing enzymes since 2004.

Typically, the Painters will introduce 900 gallons of water for every 50 bushels (approximately 2,850 pounds) of grain used. Their water source is a spring that was developed for the purpose; municipal water would have to be treated to remove chlorine that would affect the enzymes. The mash is constantly agitated while the cooking occurs. When the temperature drops and the liquefaction stage comes to an end, pH is checked and adjusted again with the use of a dilute acid or a phosphate buffer.

At this point a second series enzyme, or glucoamylase, is introduced to initiate saccharification. In the earliest days of the operation, the temperature of the vat was maintained at 120°F. Later, after some experiential fine tuning and the introduction of a different enzyme, the Painters were able to reduce that temperature to 90°F, saving that much more process energy. Temperature at this point is critical to the successful function of the enzyme.

Fermentation occurs in six 1,250-gallon stainless steel vats, so the batches can work in a six-day rotation. It takes about 3½ days for each fermentation cycle to be completed. Efficient fermentation depends on the correct dosage of yeast, based on the sugar content of the mash. Yeast is mixed and agitated briefly in a smaller vessel, then added to the mash and left to work in the sealed vat. Temperatures are held below 80°F, and anaerobic conditions are maintained through the use of a simple fermentation vapor lock to encourage the production of ethanol and CO_2. After fermentation is complete, the mixture is pumped to a separator, where the liquid beer is sent to a tank prior to the distillation column, and the

PAINTERLAND FARMS

Fig. 10.2: *A grinder/mixer includes a set of scales to weigh feedstock material.*

PAINTERLAND FARMS

Fig. 10.3: *Slurry is agitated to prepare mash for fermentation.*

Fig. 10.4: *Fermentation vats are set up in a rotation cycle so there is always fermented mash ready for distillation.*

wet solids, or distiller's grains, are collected in another container.

The 12-inch distillation column rises into the roof cupola and contains a series of perforated plates. The distillation process functions through the difference in boiling points between alcohol and water. Because of the higher vapor pressure of alcohol, ethanol evaporates faster than water when the mixture is boiled. As a result, the vapors contain a disproportionately large share of ethanol. The perforated plates in the column stage a series of evaporations as beer is added to the column. At each stage, vapors flow through the perforations in the plates and are cooled by water flowing across the plates. The ethanol rises, as vapor, and the water falls through down tubes staggered at the edge of each plate. At each level, the proportion of alcohol to water becomes a bit greater, as the process works its way to the uppermost reaches of the column.

PAINTERLAND FARMS

PAINTERLAND FARMS

Fig. 10.5: *The heating boiler and distillation column, seen in the background. Fermentation tanks are on the right.*

At the top of the column, the distilled and condensed ethanol vapors are collected and piped to a storage tank. The descended water is collected at the bottom of the tower and fed warm to the cattle as sweet water. To comply with federal regulations, the 190-proof (95 percent pure) ethanol is denatured with 2 percent unleaded gasoline to make it non-potable. It took some effort to squeeze a high-proof product out of the still, and the operators concentrated on tweaking the condensing process at the peak of the tower. The best result, according to John Sr., came about when they simply placed a ventilating fan at the top of the column to spot-cool it. The product they get is about the highest proof obtainable without utilizing special equipment, such as a molecular sieve, which is a dehydration device used industrially to remove the final 5 percent water from the mix. The sieve is made of an aluminum-silica material containing microscopic pores that only allow water and carbon dioxide to be adsorbed, leaving pure ethanol behind. This kind of high-tech investment is well beyond the farm's budget, but the Painters are confident that they may eventually be able to locate a suitable device as surplus or through an industrial auction.

Alcohol yield comes to between 110 and 120 gallons of fuel-grade ethanol per ton of grain at maximum proof. It's stored in metal and poly containers and pumped off as needed. Absolutely pure ethanol is only required for the manufacture of gasohol and other blends. The 190-proof fuel made on the farm is perfectly usable straight up in gasoline engines. The Painters use it in various farm engines on the premises as well as in farm vehicles, without blending or additional drying. The federal Alcohol Fuel Producer's permit allows for use on the farm or for travel between farm operations exclusive of a fuel excise tax.

The farm's diesel equipment is not yet on the table for ethanol use. Technicians have advised that the fuel's lack of lubricity could be problematic, even in a blend, and the working equipment is too much of an investment to risk. John Painter still plans on acquiring an older tractor or loader to proceed with some testing on his own, however.

Lower-proof fuel is regularly used in the farm's heating equipment and in the family's home furnace. It also provides space heating for cold corners of the barns and outbuildings.

By-products

For the dairy farm, the value of the by-product was a key element. Besides an immediate and significant milk production increase, the nutrient value of the soy- and distiller's grain-blended feed supplement increased markedly. The distiller's by-product consistently runs close to 26 percent protein, an 18 percent increase over corn alone, and it's a big advantage over having to buy supplemental organic soybean meal at $700 per ton.

An analysis of milk production indicated a 4 percent decrease concurrent with a decrease in the feeding of distiller's grain over a several-month period. One percent coincided with the removal of soy residue, and the remaining 3 percent drop was attributed to the final removal of corn distiller's grain from the ration.

Carbon dioxide is also a co-product of ethanol production and especially in larger operations it has marketable value. The Painters have not implemented CO_2 recovery in their operation as of yet.

A Managed Economic Cycle

The organic feature of the farm has obliged the Painters to conduct their business in a more broad and traditional manner than some operations. They do things the old-fashioned way, from breeding and raising the cattle on site to raising hay and cultivating the corn and soybeans themselves. The elder Painter says that despite the higher cost of producing alcohol from organic feedstocks, the gains from the protein-rich distiller's grains, plus the sale of feeder steers, organic hay, and high-value dairy all contribute to bringing the bottom line into balance.

Good record keeping is critical to data analysis and in estimating cost and profitability. The circumstances and results of each batch run were dutifully recorded, indicating date and time, temperatures, water volume, weight and cost (if any) of feedstock, enzyme type and cost, acid volume and cost, distiller's grain volume, and amount of ethanol and proof-strength produced.

The improvements have moved Painterland Farms closer to sustainability and self-sufficiency. Very little goes to waste in a system where recycling is crucial to making a profit. By-products are just as important as the fuel itself, and the Painters are constantly seeking to develop markets for anything with value.

Iogen Corporation: Ottawa, Ontario

Straw Into White Gold?

Iogen Corporation of Canada is a biotechnology firm engaged in the manufacture of industrial enzymes for the paper, textile and animal feed industry. Iogen is one of several companies at the forefront of developing a cellulosic ethanol process, using straws and grass-based agricultural residue feedstocks rather than the conventional food starches and sugars. Their cellulose ethanol program has been ongoing for 30 years and the company currently operates a cellulose ethanol demonstration-scale facility in Ottawa.

In March of 2008, Iogen Corporation and Sustainable Development Technology Canada (SDTC) announced that the corporation's application for funding under a Government of Canada program designed to encourage next-generation renewable fuels — the NextGen Biofuels Fund™ — had reached the due diligence phase. At the end of the whole process, it is expected that SDTC will approve the construction of a commercial-scale cellulosic ethanol biorefinery to be built in Saskatchewan.

Fig. 10.6: *Fueling up on cellulose-based E-85 ethanol at the Iogen Demonstration plant.*

The benefit of next-generation renewable fuels is that they are derived from renewable feedstocks such as agricultural residues, fast-growing grasses, and forest biomass. These sources consist mostly of materials that are viewed as waste and are grown on low-grade soil. Several grasses, among them switchgrass and an Asian variety known as miscanthus, are perennials, so they can be harvested year after year without replanting.

What's more, most cellulosic feedstocks require little, if any, fertilizers or pesticides, and they can provide more ethanol per acre, and thus greater energy yields, than corn.[1]

An Unconventional Process

Though cellulose ethanol and conventional grain-based ethanol have the same molecular makeup, the complex structure of the cell walls in wood and grass make it difficult to break down into the sugars needed for fermentation. Biomass feedstocks have an inherent structural strength that resists decomposition — and therein lies the problem. The lignin, cellulose and hemicellulose that give woody plants and stalks the ability to stand tall and in some cases achieve heights of eight feet or more are the very bane of the breakdown process.

In contrast to sugar and starch crops, this kind of biomass conversion requires additional pretreatment with heat and chemicals to expose the needed sugars. Conventional ethanol producers who use sugar crops have it easy — after mechanical chopping or milling, yeast introduced in the fermentation stage readily converts the sugars to alcohol. Even starch-based feedstocks such as corn and grains are not particularly problematic: enzymes are readily available to make the conversion, sometimes aided by heat and mild acids, prior to the fermentation stage.

But the hydrolysis process currently available to break down the chains of sugar molecules that make up cellulose prior to the enzyme treatment comes at a great cost of energy, because it relies on high temperatures, strong acids, and considerable time to achieve its goal. The method is inefficient, yet the feedstock remains viable. The potential energy stored in the cellulose and hemicellulose of some plants can comprise up to 85 percent of the plant,[2] a percentage too great to simply ignore.

Compounding the problem is the fact that conversion ratios — the percentage of ethanol achieved during fermentation — using thermochemical hydrolysis don't come close to the 15 percent or so seen in conventional starch-based conversions. The threefold difference is an economic deal-breaker in a competitive industrial market.

What Iogen and other biotech research firms such as San Diego's Verenium Corporation and Florida-based Dyadic International have done is to isolate and refine the enzymes responsible for breaking down cellulose in nature. Enzymes are complex protein molecules made up of chains of amino acids. They're found in the cells of all living creatures, and serve as catalysts for diverse chemical reactions. There are many different enzymes, each with a very specific target and each capable of functioning within a particular environment relating to pH levels, temperatures and pressure.

The search for exotic cellulose-exploiting enzymes have taken researchers all over the globe to locate organisms, fungi and bacteria that have "the right stuff" in the enzymatic sense. Costa Rican termites and certain deep-sea bacteria are promising candidates, the former for the contents of their gut, the latter for their ability to withstand extremely harsh environments. Other groups have investigated a fungus from New Guinea, *Trichoderma viride*, that is particularly adept at devouring cotton fiber.[3] Still others are looking at native microbes more suited to hydrolyzing cellulose from native plants.[4]

Yet the thing about enzymes is that they can be manufactured for commercial purposes and genetically engineered to the job at hand. Such boutique products are currently used in everything from paper production to textile manufacturing, and now hope to find a home in biofuels production. The impact of these enzymes is not incidental. The cost of enzyme conversion has been reduced thirtyfold,[5] and further gains are expected in the not-too-distant future.

The Promise of Progress

Iogen's technological process isn't necessarily innovative because of a single spectacular breakthrough — though the strain of *Saccharomyces* yeast developed at Purdue University in the mid-1990s and licensed to the biotechnology firm in 2004 [6] could certainly qualify as one. More explicitly, Iogen Corporation's accomplishments can be attributed to the way in which they've developed each step to put their process together.

Converting biomass to ethyl alcohol has never been an issue. Finding a way to make the conversion economically feasible has really been the sticking point. By examining each stage of the transformation from fiber to liquid fuel and applying well-thought strategies at every level, the firm has been able to control the things that cost money: time, energy input, and the effectiveness of the process at hand. The result affects nearly every phase of the ethanol-manufacturing process.

The sequence starts much as with any conventional ethanol operation: the plant fiber is mechanically ground to physically break it apart in preparation for pretreatment. The pretreatment step is the initial one in de-structuring the fiber and making it more accessible to enzymes by expanding its surface area through what the company describes as a "modified steam explosion process". The fiber is separated into its primary structural components, cellulose, hemicellulose and lignin. The intense treatment at this stage enhances ethanol yields and reduces aggregate cost.

Hydrolysis occurs when the enzymes are introduced. The enzymes are custom-made for the particular feedstock available, which further improves yields. A multi-stage process uses heat and chemical means to aid in the biochemical breakdown of the cellulose into its constituent sugars, glucose (a hexose, or simple sugar with six carbon atoms per molecule) and xylose (a five-carbon sugar that is difficult to ferment).

The cost of isolating the C6 sugars from the C5 xylose is prohibitive enough to price the resultant ethanol out of the competitive market with gasoline. At the same time, the rogue xylose makes up about 30 percent of available sugars in biomass and agricultural wastes, so it cannot be ignored. The recombinant yeast developed for the fermentation process simultaneously converts both types of sugars to ethanol and increases ethanol yields at the fermentation stage by up to 40 percent. The higher conversion ratios translate to competitive cost efficiencies and, at the back end, reduced waste that would otherwise have to be dealt with.

The fermented beer is distilled by conventional fractional distillation to produce the cellulose ethanol. In the end, about two thirds of the biomass feedstock is converted to ethanol, with the unfermentable remainder — mostly lignin — being separated and used as co-generative fuel to generate steam and electricity for the production process. In hard figures, that represents a yield of nearly 90 gallons of cellulose ethanol per metric ton of fiber, or about 81 gallons per US ton.[7] The energy saved by co-generation corresponds not only to a reduction in the use of conventional fossil fuels, but to a reduction in the generation of greenhouse gases due to the fact that much of the agricultural waste used in cellulose ethanol production would otherwise be burned in the fields.

In a large-scale commercial operation, heat integration, water recycling and co-product benefits all contribute to the efficiency and economical viability of the process. The opportunity for using straw stock and agricultural waste opens an entirely new class of resources, the non-food portion of renewable agricultural crops. A *Business Week* environmental report in December 2006 estimated that on the right industrial scale, "America's forests, agricultural waste, and 40 to 60 million acres of prairie grass could supply 100 billion gallons or more of fuel per year."[8] (142 billion gallons of gasoline are currently used annually.[9])

In 2006, Iogen partnered with Royal Dutch Shell and Volkswagen AG to study the feasibility of producing cellulose ethanol in Germany. That same year, the Wall Street firm Goldman-Sachs invested $30 million in the company's cellulose ethanol technology. Volkswagen sees a potential reduction in CO_2 emissions of 90 percent with cellulose ethanol as compared to conventional fuels. In Canada, where cars and trucks generate more than 30 percent of all greenhouse gas emissions, the Iogen partnership hopes to impact the transportation segment with the clean-burning fuel. And the US has more than four million Flexible Fuel vehicles in service capable of using an 85 percent ethanol/15 percent gasoline blend — though even using a 10 percent ethanol blend in conventional vehicles will reduce greenhouse gases by 16 grams of CO_2 for every mile driven.[10] Beyond greenhouse gas reductions, reliance on local sources of renewable energy offers the benefit of supporting local and regional economies, strengthening the agricultural community, and reducing dependence on petroleum imports, either by direct use or by blending the ethanol with gasoline.

Iogen Corporation

Fig. 10.7: *The evaporator at Iogen's demonstration plant.*

Living on Earth/World Media Foundation

Fig. 10.8: *Jeff Passmore of Iogen, in front of a flexible fuel vehicle powered by cellulosic ethanol.*

Genetically Modified Enzymes: The Risk?

Despite advances in isolating custom-tailored enzymes for the manufacture of cellulosic ethanol, the path to that point raises some questions. Specifically, environmental advocates are concerned that the genetic engineering responsible for the creation of these wood-eating cellulases is just another contribution to the stew of Genetically Modified Organisms (GMOs) already unleashed on the planet. Fears that these GMOs may inadvertently escape into the environment or mutate into malignant organisms remain a concern of the worldwide scientific community.

And not without reason. In one instance within the food industry, yeast DNA was inserted with multiple copies of its own genes in an effort to increase alcohol production. This unexpectedly raised levels of a naturally occurring toxin and potential carcinogen by 40 to 200 times.[11] Still, genetically modified strains of the alpha amylase enzyme used to break down starch molecules in conventional corn-based ethanol production have been in use for years.

Iogen Corporation personnel are keenly aware of the controversy and conduct their manufacturing accordingly. The production organisms they use are classified as Risk Group 0, defined as no risk to human health. Typical manufacturing involves "Good Manufacturing Practices," an industry standard that specifies careful containment and cleaning. The enzyme manufacturing is conducted in a contained vessel under aseptic conditions, required because the organisms are very specialized and not generally competitive in nature. After synthesis of the enzymes, the organisms are removed and undergo a kill step, which confirms elimination of any viable organisms.

Mandy Chepeka, the firm's spokesperson, indicated that "If Iogen's organisms escaped into the environment, they would not compete favourably on most food sources, and would predominantly be considered food for other organism [and] broken down entirely. In terms of their impact on cellulose, the organisms are not sufficiently different from the native wood rot fungus strains to act any differently in nature."

She presented Iogen's position thusly: "The specific properties of the enzymes are engineered for efficacy on a highly unnatural form of cellulose. Our pretreatment makes cellulose one-thousand times more digestible than regular fibre, and the properties engineered into the enzymes give advantage under high concentrations of product and high temperatures, not those ever found in nature. So, we would expect the inadvertent or accidental release of organisms to the environment to be low impact, no different from a regular wood rot fungus."

Conclusion: What Next?

Now that you've arrived at the end of *Alcohol Fuel*, I have probably either roused your interest or estranged you from the prospect entirely. Before we leave the topic altogether, though, I'd like to outline a few arguments in defense of fuel alcohol, for what it's worth. There seem to be four overarching reasons why people are motivated to make their own ethanol fuel, and perhaps you can find yourself in one of them.

Economic incentives: Using waste and spoilage to make your own fuel can save you a considerable amount of money, depending upon how you value your labor and how astute you are about sourcing feedstocks. The value of good judgment also extends to acquiring equipment, as anything adapted from salvage or fabricated yourself is going to be considerably less expensive than purchased apparatus.

The fuel cost per gallon is always subjective, though, because no two manufacturing operations are alike. I cringe when I see websites and promotions that lure people with alcohol prices of $0.43 per gallon or whatever, because I know how unrealistic that is for a long-term objective. I wouldn't count on making a financial killing using alcohol fuel — but, as gasoline approaches the $4.25 per gallon level (as it did while this book was being written), alcohol becomes very competitive, especially as regular petroleum supplies become scarce (and they did on several occasions in the same time period).

Your energy balance, or Energy Return on Investment (EROI) is part of the equation, and that's why I support the use of renewables such as wood fuel or solar energy, where practicable, for heating or preheating. Energy balance is the relationship between the amount of energy required to extract a fuel source and the amount of energy contained in it. On a full life-cycle scale, this can get very complicated, but in the microcosm of your production facility, it's fairly simple to calculate: If it costs you more to collect, process, heat and distill your feedstock than you get back in fuel volume, you're on the losing side of the equation. Since energy consumed in preparation and distillation is such an important facet, you'd be wise to pay it some mind.

Don't forget, either, that there are federal tax credits available for the small ethanol producer, encapsulated in the Small Producer Tax Credit, which allows a credit of $0.10 per gallon for the first 15 million gallons produced — which would cover most people. You can check with your state and provincial governments as well for any renewable energy tax credits, grants or low-interest loans that might be available in your own jurisdiction.

Environmental grounds: Certainly more than a few people would look to alcohol fuel as a means to reduce automotive and internal-combustion emissions. The sharp reduction in the amount of hydrocarbons and carbon dioxide at the tailpipe is a real incentive to taking ethanol fuel seriously. Critics can point to questions of issues over oxides of nitrogen and aldehydes in the fuel, but the reality is that control of combustion chamber temperatures and catalytic combustor technology is fully accessible, and at this point it is more an accounting matter than an engineering one. There is no petroleum-based fuel and arguably few renewable fuels that are as clean burning as alcohol.

Technically, the Environmental Protection Agency prohibits tampering with the fuel and emissions systems of any highway vehicle manufactured since the early 1970s, based on the assumption that a consumer could compromise the effectiveness of the manufacturer's components. Although an exemption is available for aftermarket manufacturers to supply kits and modification equipment, the testing requirements are often prohibitively expensive and only a handful of firms pursue it. I am not aware of any case in which the EPA has pursued an individual for tampering.

Growing or recycling your own crop or feedstock is far more environmentally sound than pumping petroleum out of the ground and refining it. The ultimate solar collector, after all, is the soil in which our plants, crops and trees grow. It's difficult to argue with a system which has worked pretty well, all things considered, since the beginning of time.

Personal satisfaction: For certain people, there is no joy greater than tinkering with mechanical things. At once, it offers a sort of relaxation and a sense of accomplishment that cannot be duplicated for love or money. Building distillation equipment and messing around under the hood of a car is the perfect pastime for that type of person. What's more, even computer and electronic geeks can find a home here. The control systems used in automated distillation and the microprocessor control units integrated into modern automobiles are a beacon for creative types, begging to be modified. Even those not mechanically or electronically inclined might find contentment in the cooking aspects of alcohol production. Fermentation is, after all, a recipe, as any amateur vintner or beermaker can attest to.

Independence from suppliers: Let's not forget that we are in interesting times. The volatility of fuel prices and supplies is not just a footnote in some economist's academic report; it is a fact of life that has actually materialized in the marketplace. I'm not an economist, but I'd predict that things are going to get worse before they get better, and those who are prepared with some viable alternatives are going to be far ahead of those who've adopted a wait-and-see approach. These types of events tend to happen very quickly once they're initiated.

Appendix A: True Percentage of Proof Spirit

The table below is an abridged version of Section 30.61 of the Alcohol and Tobacco Tax and Trade Bureau's Distilled Spirits Gauging Manual, "Table No. 1: The True Percents of Proof Spirit for any Indication of the Hydrometer at Temperatures between 0° and 100°F." It is available online in its entirety at ttb.gov/foia/Gauging_Manual_Tables/Table_1.pdf.

This table, abridged due to limitations of space, shows the true percent of proof of distilled spirits for hydrometer readings between 151 and 206 at selected temperatures between 61° and 100° F. The column at left shows the hydrometer reading, and on the same horizontal line, the corrected reading or "true percent of proof" corresponding to the temperature readings listed in the top column. The table is computed for tenths of a percent. For actual readings at temperatures not listed, consult the online reference for accuracy.

Example:
Hydrometer reading 192
Temperature, ° F 70
True percent of proof 189.6

Hydrometer Reading	Temperature °F								
	61	65	70	75	80	85	90	95	100
151	150.7	149.3	147.5	145.7	143.9	142.1	140.3	138.4	136.5
152	151.7	150.3	148.5	146.8	145.0	143.2	141.4	139.5	137.6
153	152.7	151.3	149.5	147.8	147.0	144.2	142.4	140.5	138.7
154	153.7	152.2	150.6	148.8	148.1	145.2	143.4	141.6	139.7
155	154.7	153.3	151.6	149.8	148.1	146.3	144.5	142.6	140.8
156	155.7	154.3	152.6	150.9	149.1	147.3	145.5	143.7	141.8
157	156.7	155.3	153.6	151.9	150.1	148.4	146.6	144.7	142.9
158	157.7	156.3	154.6	152.9	151.2	149.4	147.6	145.8	144.0
159	158.7	157.3	155.6	153.9	152.2	150.4	148.7	146.9	145.0

Hydrometer Reading	Temperature °F								
	61	65	70	75	80	85	90	95	100
160	159.7	158.3	156.6	154.9	153.2	151.5	149.7	147.9	146.1
161	160.7	159.3	157.7	156.0	154.3	152.5	150.8	149.0	147.1
162	161.7	160.4	158.7	157.0	155.3	153.6	151.8	150.0	148.2
163	162.7	161.4	159.7	158.0	156.3	154.6	152.8	151.1	149.3
164	163.7	162.4	160.7	159.0	157.4	155.6	153.9	152.1	151.4
165	164.7	163.4	161.7	160.1	158.4	156.7	154.9	153.2	151.4
166	165.7	164.4	162.8	161.1	159.4	157.7	156.0	154.2	152.5
167	166.7	165.4	163.8	162.1	160.5	158.8	157.0	155.3	153.5
168	167.7	166.4	164.8	163.2	161.5	159.8	158.1	156.4	154.6
169	168.7	167.4	165.8	164.2	162.5	160.9	159.2	157.5	155.7
170	169.7	168.4	166.8	165.2	163.6	161.9	160.2	158.5	156.8
171	170.7	169.4	167.9	166.2	164.6	163.0	161.3	159.6	157.8
172	171.7	170.4	168.9	167.3	165.7	164.0	162.3	160.6	158.9
173	172.7	171.5	169.9	168.3	166.7	165.1	163.4	161.7	160.0
174	173.7	172.5	170.9	169.3	167.7	166.1	164.5	162.8	161.1
175	174.7	173.5	172.0	170.4	168.8	167.2	165.5	163.9	162.2
176	175.7	174.5	173.0	171.4	169.8	168.3	166.6	165.0	163.3
177	176.7	175.5	174.0	172.5	170.9	169.3	167.7	166.1	164.4
178	177.7	176.5	175.0	173.5	172.0	170.4	168.8	167.2	165.5
179	178.7	177.5	176.1	174.6	173.0	171.5	169.9	168.3	166.6
180	179.7	178.6	177.1	175.6	174.1	172.5	171.0	169.4	167.7
181	180.7	179.6	178.1	176.7	175.1	173.6	172.1	170.5	168.8
182	181.7	180.6	179.2	177.7	176.2	174.7	173.2	171.6	170.0
183	182.7	181.6	180.2	178.8	177.3	175.8	174.3	172.7	171.1
184	183.7	182.6	181.3	179.8	178.4	176.9	175.4	173.8	172.2
185	184.7	183.7	182.3	180.9	179.4	178.0	176.5	174.9	173.4
186	185.7	184.7	183.3	181.9	180.5	179.1	177.6	176.1	174.5
187	186.7	185.7	184.4	183.0	181.6	180.2	178.7	177.2	175.7
188	187.7	186.7	185.4	184.1	182.7	181.3	179.8	178.4	176.8
189	188.8	187.8	186.5	185.2	183.8	182.4	181.0	179.5	178.0
190	189.8	188.8	187.5	186.2	184.9	183.5	182.1	180.7	179.2
191	190.8	189.8	188.6	187.3	186.0	184.6	183.2	181.8	180.3
192	191.8	190.8	189.6	188.4	187.1	185.8	184.4	183.0	181.5
193	192.8	191.9	190.7	189.5	188.2	186.9	185.5	184.1	182.7
194	193.8	192.9	191.7	190.5	189.3	188.0	186.7	185.3	183.9

Hydrometer Reading	Temperature °F								
	61	65	70	75	80	85	90	95	100
195	194.8	193.9	192.8	191.6	190.4	189.2	187.8	186.5	185.1
196	195.8	195.0	193.9	192.7	191.5	190.3	189.0	187.7	186.4
197	196.8	196.0	194.9	193.8	192.7	191.4	190.2	188.9	187.6
198	197.8	197.0	196.0	194.6	193.8	192.6	191.4	190.1	188.9
199	198.8	198.1	197.1	196.0	195.0	193.8	192.6	191.4	190.2
200	199.8	199.1	198.1	197.1	196.1	195.1	193.9	192.8	191.6
201	—	—	199.2	198.2	197.3	196.3	195.2	194.1	192.9
202	—	—	—	199.3	198.5	197.5	196.5	195.4	194.2
203	—	—	—	—	199.7	198.7	197.8	196.7	195.6
204	—	—	—	—	—	200.0	199.1	198.0	196.9
205	—	—	—	—	—	—	—	199.3	198.3
206	—	—	—	—	—	—	—	—	199.6

Appendix B: Resources and Suppliers

Pumps, Valves, Handling and Fabrication Equipment

Retail and commercial suppliers of plumbing parts, filters and equipment

W.W. Grainger, Inc.
(800) 323-0620
grainger.com

McMaster-Carr
P.O. Box 740100
Atlanta, GA 30374-0100
(404) 346-7000
mcmaster.com

Northern Tool & Equipment
2800 Southcross Drive West
Burnsville, MN 55306
(952) 894-9510
northerntool.com

Tractor Supply Co.
200 Powell Place
Brentwood, TN 37027
(877) 718-6750
tractorsupply.com

Column (Tower) Packing

Pall and Raschig Rings, Saddle Packing

Amistco Separation Products, Inc.
23147 Highway 6
Alvin, TX 77512
(281) 331-5956
amistco.com

Newsmart (Aust) Pty, Ltd.
4/53-57 West Street
Hurstville 2220 NSW Australia
02 9586 3635
conveyor-machine.com

Smiley's Home Distilling
2 Cyrus Court
Nepean, ON K2H 9C9 Canada
(613) 820-1069
home-distilling.com

The Pall Ring Company, Ltd.
New Road, Crimplesham
Kings Lynn, Norfolk PE33 9ES UK
+44(0) 1366 389680
pallrings.co.uk

ACS Industries, LP
14211 Industry St.
Houston, TX 77053
(800) 231-0077
acsseparations.com

Adsorbents and Molecular Sieves

Industrial suppliers of drying materials and equipment

Delta Adsorbents
24 Congress Circle
Roselle, IL 60172
(800) 274-3205
deltaadsorbents.com

AGM Container Controls, Inc.
3526 East Fort Lowell Road
Tucson, AZ 85716
(800) 995-5590
agmcontainer.com

Hengye USA
1241 Ellis Street
Bensenville, IL 60106
(800) 683-0464
hengyeusa.com

Enzymes and Yeasts

Manufacturers and industrial suppliers may refer queries to other outlets for individual sale.

Novozymes North America, Inc.
77 Perry Chapel Church Road
Franklinton, NC 27525
(919) 494-3000
novozymes.com/en/mainstructure/

Danisco US, Inc., Genencor Division
200 Meridian Centre Blvd.
Rochester, NY 14618
(800) 847-5311
www.genencor.com

Murphy & Son, Ltd.
Alpine Street
Old Basford, Nottingham, NG6 0HQ
0115 978 0111
murphyandson.com

Maps (India), Limited
302, Shapath-3,
Near GNFC Info Tower, S.G. Road,
Ahmedabad 380054 India
+91 (79) 26859971-74
mapsenzymes.com

CBS Chemicals and Biochemicals Supplier
Rue du cloître n. 75 - B-1020 Brussels
+32 2 478 56 18 Belgium
cbs.brew.com

Specialty Enzymes and Biochemicals Co.
13591 Yorba Ave.
Chino, CA 91710
(909) 613-1660
specialtyenzymes.com

Brewers Supply Group West
3624 Munster Ave., Units D-E
Hayward, CA 94545
(800) 374-2739
brewerssupplygroup.com

Brew Supplies.com
Sumac Enterprises
181 County Road 2220
Cleveland, TX 77327
(281) 432-1150
brewsupplies.com

Enzyme Development Corp.
21 Penn Plaza
360 West 31st Street
New York, NY 10001
(212) 736-1580
enzymedevelopment.com

Automotive and Aftermarket Suppliers

Engine components and performance modifications

The Eastwood Company
263 Shoemaker Road

Pottstown, PA 19464
(800) 544-5118
eastwoodco.com

J.C. Whitney
761 Progress Parkway
LaSalle, IL 61301
(800) 603-4383
jcwhitney.com

Holley Performance Products
1801 Russellville Road
Bowling Green, KY 42101
(270) 782-2900
holley.com

NAPA Auto Parts
Genuine Parts Company, Headquarters
2999 Circle 75 Parkway Southeast
Atlanta, GA 30339-3050
(877) 805-6272
napa.com

Alcohol Fuel Permits and Regulations

Contact state alcohol enforcement or revenue departments for individual state requirements.

Alcohol and Tobacco Tax and Trade Bureau
Public Information Officer
1310 G Street, NW
Suite 300, Washington, DC 20220
(202) 927-5000
ttb.gov

Laboratory and Testing Equipment

Hydrometers, refractometers, saccharometers, thermometers, pH meters and supplies

Nova-Tech International, Inc.
800 Rockmead Drive, Suite 102
Houston, TX 77339-2496
(866) 433-6682
novatech-usa.com

VWR Labshop
800 East Fayban Parkway
Batavia, IL 60510
(866) 360-7522
vwrlabshop.com

Fisher Scientific
2000 Park Lane Drive
Pittsburgh, PA 15275
(800) 766-7000
fishersci.com

ScienceLab
14025 Smith Road
Houston, TX 77396
(800) 901-7247
sciencelab.com

Utech Products, Inc.
135 Broadway
Schenectady, NY 12305
(518) 489-5705
utechproducts.com

Cole-Parmer Instrument Company
625 East Bunker Court
Vernon Hills, IL 60061-1844
(800) 323-4340
coleparmer.com

Cole-Parmer Canada, Inc.
210 - 5101 Buchan St
Montreal, QC H4P 2R9 Canada
(800) 363-5900
coleparmer.ca

Chemical Suppliers

Acids, alkalines, buffers, additives

Southern States
P.O. Box 26234
Richmond, VA 23260-6234
(804) 281-1000
southernstates.com

Bio-WORLD
PO Box 888
Dublin, OH 43017
(800) 860-9729
bio-world.com

Advance Scientific & Chemical, Inc.
2345 SW 34th Street
Fort Lauderdale, FL 33312
(800) 524-2436
advance-scientific.net

Safety and Handling Equipment

Respirators, eye protection, safety clothing and equipment

Lab Safety Supply, Inc.
P.O. Box 1368
Janesville, WI 53547-1368
(800) 356-0783
labsafety.com

Northern Safety Co., Inc.
P.O. Box 4250
Utica, NY 13504-4250
(800) 571-4646
northernsafety.com

Organizations and Associations

Renewable fuels, agriculture, specialty trade groups or organizations

National Ethanol Vehicle Coalition
3216 Emerald Lane, Suite C
Jefferson City, MO 65109
(573) 635-8445
e85fuel.com

American Coalition for Ethanol
5000 S. Broadband Lane, Suite 224
Sioux Falls, SD 57108
(605) 334-3381
ethanol.org

Canadian Renewable Fuels Association
350 Sparks Street, Suite 1005
Ottawa, ON K1R 7S8 Canada
(613) 594-5528
greenfuels.org

Enzyme Technical Association
1601 K Street, NW
Second Floor
Washington, DC, 20006
enzymetechnicalassoc.org

Ethanol Producers and Consumers
172 Ball Road
Nashua, MT 59248
(406) 785-3722
ethanolmt.org

Sustainable Agriculture Research and Education
USDA-CSREES
Stop 2223
1400 Independence Ave. SW
Washington, DC 20250-2223
(202) 720-5384
sare.org

ATTRA — National Sustainable Agriculture Information Service
P.O. Box 3657
Fayetteville, AR 72702
attra.org

Courier-service National Laboratories

National Testing Laboratories, Ltd.
6571 Wilson Mills Road
Cleveland, OH 44143
(800) 458-3330

Suburban Water Testing Labs
4600 Kutztown Road
Temple, PA 19560
(800) 433-6595
h2otest.com

Glossary of Terms

Absorption: The penetration of a substance into the bulk of a solid or liquid through permeation.

Adsorption: The surface retention of solid, liquid or gas molecules by a solid or a liquid.

Aerobic: Able to survive or grow only in the presence of air or free oxygen. Organisms such as yeast can live in aerobic and anerobic (airless) conditions.

Alpha amylase: An enzyme found in animal, plant, fungal and bacterial sources. It hydrolyzes starch into intermediate dextrin sugars.

Amorphous: Materials without a crystalline structure.

Amylase: A class of enzymes that hydrolyzes starch into sugars. Plants that store their sugars in the form of starch manufacture amylase enzymes, as do many bacteria and fungi.

Amyloglucosidase: The enzyme that hydrolyzes starch to glucose.

Amylopectin: The outer insoluble part of a starch granule.

Amylose: The inner soluble portion of a starch granule.

Anaerobic: Able to survive or grow in an environment free of oxygen. During fermentation, when available oxygen is used up, yeast changes its metabolism to function in an anaerobic state to produce ethyl alcohol and carbon dioxide.

Autoignition temperature: The lowest temperature at which a liquid's vapors will self-ignite, without direct exposure to spark or flame.

Azeotrope: A mixture of two or more liquids that behaves as one with a constant boiling point. When the ethanol content in water reaches 95.57 percent, an azeotrope forms, and the two substances can no longer be separated by simple distillation.

Azeotropic distillation: A distillation technique that uses a ternary or third element such as benzene or trichloroethylene to bond with the remaining water so pure alcohol can be drawn off separately.

Backslop: Spent mash, or the liquid remaining after distillation, which contains a mixture of dead yeast and unfermentable materials.

Bagasse: The fibrous cane byproduct of sugar juice extraction in the production of ethyl alcohol or sugar. Once dried, it is used as a biomass fuel.

Barley malt: Barley grains prepared for the purpose of making enzymes through germination and drying. The enzymes produced include alpha and beta amylase needed for making alcohol.

Batch distillation: The simplest method of distillation, in which a single load of fermented mash is run through the still and removed after the anticipated proof is obtained. The still is reloaded with fresh mash for each subsequent run.

Beer: The alcohol-laden liquid produced in the mash-fermenting process. For clarification, the liquefied feedstock is generally called mash. More specifically, it is called "wort" before the introduction of yeast, and "beer" after yeast has been added.

Benzene: A highly carcinogenic liquid hydrocarbon sourced from coal or petroleum. Used as an additive to raise the octane rating of gasoline and used in azeotropic distillation.

Beta Amylase: An enzyme that hydrolyzes starch into simple sugars. The beta amylase enzyme is found in higher plant life; it converts dextrins to maltose, a disaccharide, containing two units of sugar.

Biomass: A source of fuel derived from any living organism, except for fossil fuels. These would include agricultural crop residues, forest products, wood and animal wastes, plants and plant residues, forages and municipal wastes.

Boiling point: The temperature at which a liquid turns to vapor, dependent upon atmospheric pressure.

Boiler tank: The vessel in a simple pot still which contains heated mash for distillation.

Bottoms: The material remaining at the bottom of the distillation column after the alcohol has been distilled out.

Bottom dead center (BDC): The lowest point of the piston travel in an internal combustion engine, when the crank is at the bottom of its stroke.

British Thermal Unit (Btu): A unit of heat energy equal to the amount of heat required to raise the temperature of one pound of water one degree Fahrenheit at 60°F and one standard atmosphere of pressure. The equivalent of 252 calories or 1,054 joules.

Brix scale hydrometer: A saccharometer, or hydrometer delineated in Brix scale, used to measure the specific gravity of a solution containing sugar.

Bubble cap column: A traditional distillation column design which uses equilibrium-stage plates with vented and capped risers to transfer vapors through liquid from one stage to the next. This design operates at low vapor and liquid rates.

Bureau of Alcohol, Tobacco and Firearms (BATF): The Federal government agency formerly responsible for permitting for the production of fuel and beverage alcohol. In 2003, the task was transferred to the Treasury Department's Alcohol and Tobacco Trade and Tax Bureau, or TTB.

Calorie: A unit of heat energy, equal to 4.18 joules. It is the amount of heat required to raise the temperature of one gram of water by one degree Centigrade at a pressure of one standard atmosphere.

Carbohydrate: A compound consisting of carbon, hydrogen and oxygen. The most abundant of all organic compounds, includes starches, sugars and cellulose.

Carbon dioxide (CO_2:) A colorless gas produced by burning, decomposing or breathing carbon and organic compounds.

Carbon monoxide (CO): A colorless, flammable and poisonous gas produced by the incomplete combustion of carbon or a carbohydrate.

Carburetor: A mixing apparatus in an internal combustion engine that meters fuel with air, at a ratio depending upon engine requirements.

Casein: The protein portion of milk.

Catalyst: In ethanol production, a form of protein that can initiate a biochemical reaction without being altered or destroyed itself.

Catalytic converter: A component of a vehicle's exhaust system that combusts unburned hydrocarbons and oxides of nitrogen at lower temperatures than they would normally, to reduce harmful emissions.

Cellulase: An enzyme that breaks down cellulosic material such as wood and plant fibers into glucose.

Cellulose: The major component of wood and vegetable fibers and plant cell walls.

Centrifugal pump: A rotary pump that moves liquid by rotary or centrifugal force.

Cetane: A measurement used for diesel fuel.

Chemurgy: A discipline devoted to the development of chemical and industrial products from organic and agricultural sources.

Closed loop mode: The term used when a modern microprocessor-controlled vehicle is functioning within the parameters of its oxygen sensor to monitor and regulate fuel mixture and other factors.

Co-generation: The concurrent use of two types of energy from a single fuel source. Thermal energy harvested from a diesel- or gasoline-powered electrical generator is a common co-generation practice.

Column: A vertical fractionating column used in the distillation of fuel, beverage or industrial alcohol.

Combustion chamber: In an automotive engine, the cavity formed between the cylinder head and piston where ignition and combustion of the fuel-air mixture occurs.

Common rail: A component of a fuel injection system that supplies the individual fuel lines to each fuel injector in an internal-combustion engine.

Compound: A combination of two or more elements whose molecules consist of unlike atoms.

Compression ratio: In an internal-combustion engine, the measure of the proportion of volume in a fully expanded piston compared to the volume when it is fully compressed. Expressed in ratio form, with the first term the larger of the two, e.g. 8.5 :1.

Compression-ignition engine: A diesel engine, or one that uses the heat of compression to ignite the air-fuel mixture rather than the ignition of a spark.

Complex sugars: Sugar composed of three or more monosaccharides, or single sugar units. Complex carbohydrates.

Condensed distiller's solubles (CDS): Thin stillage from the distillation process that is evaporated to create a viscous syrup used for animal feed.

Condenser: A piece of equipment that cools, or removes heat, from a vapor so it will condense to a liquid.

Continuous distillation: A distillation process in which fermented mash is continuously fed to the still and alcohol extraction is ongoing without interruption at the desired proof.

Control coil: A heat exchanger positioned at a specific point within a distillation column to control temperatures at that point.

Conversion: A step in the fermentation process in which glucoamylase enzymes reduce complex sugars to simple fermentable sugars. Also called the saccharification step.

Cooker: The vat or tank which heats and agitates the mash solution for fermentation.

Cooling coil: A liquid-to-liquid heat exchanger immersed in a tank for purposes of cooling.

Co-products: Products and substances other than ethanol which are created during the manufacture of alcohol fuel.

Counter-current stream: The opposing flow of materials in alcohol distillation in which ethanol vapors rise and water falls, necessary to increase the proof strength of the product.

Dehydration: The process of removing moisture from a material to dry it. Also a process in the production of anhydrous alcohol in which the slight bit of remaining moisture is removed from conventionally distilled alcohol.

Dewater: To remove water from a feedstock or from residual animal feed after distillation.

Denaturant: A substance such as gasoline, kerosene or naptha added in small quantities to fuel ethanol to make it unfit for human consumption.

Dextrin: A semi-complex sugar material hydrolyzed from starch that must be further broken down to a simpler malt sugar, maltose and glucose before being consumed by yeast.

Diastase: An enzyme that catalyzes the hydrolysis of starch to maltose.

Diaphragm pump: A positive displacement pump that uses a flexible diaphragm or membrane in place of a piston to better handle slurries and solid material.

Diffusion: The transition of dissolved solids from areas of high concentration to areas of low concentration.

Direct refluxing: A process by which droplets of alcohol are introduced into a distillation column to control temperature and increase proof strength.

Disaccharide: A sugar compound made up of two monosaccharide units.

Displacement: The volume of air-fuel mixture in the total of all of an engine's cylinders, expressed in cubic centimeters (cc's), cubic liters or cubic inches (CID).

Distillate: The part of a liquid that is extracted as vapor and condensed in the distillation process.

Distillation: The process of separating a specific vapor from a liquid mixture by the use of heat and the differences in boiling points of the substances in the mixture, then condensing and collecting the vapor as a liquid product.

Distiller's dried grains (DDG): The protein-rich granular solids remaining in mash after distillation and used as animal feed after being separated and dried.

Distiller's dried grains with solubles (DDGS): A combination of distiller's dried grains and the thin stillage called distiller's solubles which contains fine particles and proteins with nutrients.

Distiller's dried solubles (DDS): Distiller's solubles that are evaporated and condensed to supplement animal feed, or mixed with distiller's dried grains to make DDGS.

Distiller's Solubles (DS): The liquid portion of the mash left over after the distiller's grains are removed and dried. Also known as thin stillage, it contains fine protein-rich particles and nutrients and is evaporated to supplement animal feed.

Distillery: A manufacturing plant of any size containing all the components necessary to manufacture alcohol. The term "still" is a derivative of the word but describes only the distillation column and any associated mash containers and condensers.

Distributor: In a spark-ignition engine, the mechanism that delivers high-voltage electricity to the spark plugs in the correct timing sequence. In a compression-ignition engine, a fuel-delivery device connected to the injectors.

Downcomer: In a plate or tray distillation column, a conduit designed to drain liquid from one plate to the plate directly below it to promote fractionation.

Dry-milling: A process in which the whole grain seed is ground in preparation for fermentation, and co-products are separated out afterwards.

E-85: An automotive fuel consisting of 85 percent ethanol and 15 percent unleaded gasoline, developed for use in Flexible Fuel vehicles (FFVs) marketed by most major auto manufacturers.

E-95: An automotive fuel consisting of 95 percent ethanol and 5 percent gasoline, used in specially modified diesel engines in the US and Europe.

Electronic control unit (ECU): The microprocessor device used in modern electronically controlled vehicles to assess a variety of engine and environmental conditions and moderate the engine components' response accordingly. Also called a microprocessor control unit (MCU) by some manufacturers.

Electronic fuel injection(EFI): An electronically controlled fuel system that delivers a precise amount of fuel to each cylinder through a high-pressure injector.

Embden-Meyerhof Cycle: The process of fermentation, by which glucose is consumed by yeast, and a variety of enzymes work on the sugar molecule to convert it to ethyl alcohol and carbon dioxide. Named after the research of physiological chemist Gustav Georg Embden and biochemist Otto Meyerhof.

Energy balance: The proportional relationship between the amount of energy required to extract a resource and the amount of energy contained in that resource.

Enrichment: The process of raising the proof strength of a distilled product by condensing its vapors and revaporizing them.

Enzyme: A protein biocatalyst manufactured by living cells of plants and animals that promote chemical reactions without being destroyed or altered.

Equilibrium: The condition in a distillation column when the change of liquid-to-vapor and vapor-to-liquid achieves a level of balance, with regard to the exchange of molecules. In a plate or tray distillation column, all plates or phases must be at the correct temperature to achieve equilibrium. Also known as continuous equilibrium vaporization.

Ethanol: The common name for ethyl alcohol.

Ethyl alcohol: A colorless, flammable, hygroscopic liquid miscible with water and other alcohols, fermented from starch and sugar feedstocks. Also known as ethanol or grain alcohol.

Evaporation: The change of liquid to vapor by the addition of latent heat.

Exhaust gas recirculation (EGR)**:** A practice used in internal combustion engines to reduce oxides of nitrogen emissions. A portion of the engine's exhaust gases are reintroduced to the combustion chamber through the intake manifold to lower peak combustion temperatures.

Exhaust gas temperature (EGT)**:** The temperature of combusted gases exiting the combustion chamber of an engine, and used to determine proper air-fuel ratios and combustion efficiency.

Feedback carburetor: A fuel delivery device employed by auto manufacturers in the transitional period between the use of conventional carburetion and the adoption of electronic fuel injection. These carburetors regulate air-fuel mixtures based on feedback from various sensors on the engine.

Feedstock: The raw material or substrate that ultimately provides sugar for fermentation.

Fermentation: The chemical process brought about by the action of yeast and enzymes to convert fermentable sugars to alcohol, with carbon dioxide and heat as co-products.

Fermentation lock: A water-sealed device used to release carbon dioxide in the fermentation process while excluding free oxygen and bacteria.

Fermentation tank: A vat or vessel used to turn fermentable sugars in a mash to alcohol with the addition of yeast.

Firebox: The heating chamber for a cooker or boiler when solid fuel such as wood is used as a fuel source.

Flame propagation: The outward spread of flame from the original point of combustion.

Flash point: The lowest temperature at which vapors from a volatile liquid will ignite on application of a flame under stated conditions.

Flexible Fuel vehicle (FFV)**:** A factory-manufactured vehicle capable of operating on E-85 ethanol and unleaded gasoline, and any combination of blends of the two. Also, any vehicle modified to operate on two or more fuels or combinations thereof.

Flinty starch: A "hard" starch, as in the type found in corn, that is encased in a difficult-to-digest protein and requires additional heat or enzyme preparation prior to fermentation.

Fractionation: Separation of a mixture in successive stages, each stage removing some portion of the mixture, based on boiling points.

Fractional distillation: The separation of a mixture into components with different boiling points. The components are collected as fractions of the original mixture through vaporization and condensation.

Fructose: A simple sugar or monosaccharide present in fruits, nectar and honey.

Fuel injector: A device with a nozzle and control valve that delivers a precise amount of fuel to each cylinder under conditions of high pressure.

Fusel oils: An oily mixture of toxic amyl alcohols representing an insignificant fractional percentage of total alcohol output in distillation.

Gasohol: A mixture of ethanol and unleaded gasoline sold commercially at fuel pumps in many states, the ethanol content ranging between 5 and 15 percent of total volume.

Gate valve: A plumbing valve designed for maximum flow when opened, consisting of a gate or door within an orifice.

Gelatinization: The rupturing of a starch cell, which forms a gel that is difficult to agitate and can ultimately limit the hydrolysis of starch to dextrins, even with heat and enzymes present. Too-rapid gelatinization squanders time because the heating and agitation process must be temporarily suspended to allow enzymes to work on hydrolysis of the available materials.

Genetically modified organism (GMO): An organism which has had its genetic makeup altered by combining DNA molecules from dissimilar sources to produce certain characteristics, such as freeze-resistance or high sugar content in crops.

Germination: The process a seed goes through when coming out of dormancy, resulting in the sprouting of the stalk and rootlet, and the activation of enzymes.

Glow plug: An electric heating element used as a supplemental aid in cold-starting compression-ignition (diesel) engines.

Glucoamylase: An enzyme derived from fungi, capable of hydrolyzing starch completely into a fermentable glucose sugar. It is a long-term saccharifying enzyme because it works along the entire starch chain and hydrolyzes one unit a time.

Glucose: The most common monosaccharide, or simple sugar, found in many carbohydrates. Fructose and mannose (exuded as manna from the flowering ash) are two other monosaccharides.

Greenhouse gas: One of several gases such as carbon dioxide that contribute to the greenhouse effect by trapping heat from the sun.

Gristmill: A device for grinding or milling grain into grist for feed or meal.

Hammermill: An impact machine that reduces the size of materials by tearing, fracturing or crushing them.

Heat exchanger: A device that transfers heat from one medium to another, or between similar mediums, while keeping them separated.

Heating value: The measure of heat released in combustion under fixed conditions at a constant pressure. Also known as the Heat of Combustion.

Hydrocarbon (HC): A chemical compound of hydrogen and carbon, and the major component of petroleum and fossils fuels.

Hydrolysis: A chemical process in which water reacts with another substance to form new substances. Feedstock substrates are hydrolyzed, or broken down, during the cooking process using heat and water in the presence of enzymes.

Hydrometer: An instrument used to measure the specific gravity of liquids, including ethanol. The higher the specific gravity, the more buoyant the hydrometer in the liquid.

Hydrophilic: Having an affinity for water, or dissolving in, adsorbing or attracting water.

Hygroscopic: Having an ability to speed the condensation of water vapor, or absorb moisture from the air.

Idle circuit: The bypass system in a carburetor that allows the engine to receive the appropriate mixture of air and fuel for operation while the throttle is closed.

Ignition timing: The synchronization of spark or ignition with the position of the piston in a spark-ignited internal combustion engine.

Inches of mercury: A unit of measure for atmospheric pressure, used when measuring vacuum draw. Taken from the practice of using a graduated column of the element mercury at a fixed temperature to determine pressure.

Internal combustion engine: A machine or mechanism that produces work by combusting a mixture of fuel and air internally.

Inulin: A polysaccharide carbohydrate found in the roots of certain plants.

Invertase: An enzyme that converts sucrose to the monosaccharides glucose and fructose. Also known as saccharase.

Joule: A measure of the amount of work accomplished or energy consumed. Used to measure thermal, electrical and mechanical energy in the meter-kilogram-second system of units.

Kilowatt-hour: A unit of electrical energy equal to 1,000 watt-hours.

Knock sensor: An auditory instrument that detects the peculiar frequency of preignition or knocking within the combustion chamber of an engine and transmits a signal to the microprocessor control unit which adjusts timing accordingly.

Lactic acid: An acidic chemical compound used to lower the pH level of a mash mixture during fermentation. Manufactured synthetically and also formed by bacterial fermentation.

Lactose: A milk sugar derived from whey and fermented by the yeast *Saccharomyces fragilis.*

Laminar flow: A streamline flow of liquid and vapor in distinct and separate channels, undesirable in packed distillation columns.

Latent heat of vaporization: The amount of energy needed to change a liquid to a vapor, or evaporate a unit mass of liquid at a constant temperature and pressure.

Lean mixture: A fuel-air mixture in an internal combustion engine that contains more air or less fuel than the stoichiometric ideal.

Lignin: A polymer that, together with cellulose, forms the woody cell walls of plants and binds them together.

Liquefaction: The conversion, during the fermentation process, of starch to water-soluble dextrin sugar carbohydrates.

Low wines: The final portion of a batch distillation run in which the last 20 percent or so of the alcohol is removed without refluxing, and recycled with other low wines in a separate distillation run.

Lysine: An amino acid obtained from proteins and an important constituent of animal feeds. A component of distiller's dried grains with solubles.

Malt: Barley grains that have been artificially germinated for the purpose of producing enzymes, including alpha and beta amylase needed for making alcohol.

Maltase: An enzyme which breaks down maltose sugar to glucose, a simple monosaccharide. Also known as alpha-glucosidase.

Maltose: Malt sugar, or a disaccharide sugar containing two molecules of glucose, produced when starch is broken down with beta-amylase enzymes.

Manifold: A branch pipe device in an internal combustion engine that delivers vapors or gases from a single point source to multiple ports, or from multiple ports to a single terminus. Also applies to single-cylinder applications in which one port is involved.

Mash: Traditionally, the grain and water mixture made for fermentation, once yeast has been added. Prior to the introduction of yeast, the mixture is a wort; after fermentation, it is a beer.

Mash pot: The tank assembly of a batch distillery, where the fermented mash, or beer, is boiled for distillation in the column.

Mass airflow sensor: A device which calculates the density of air entering the intake manifold and provides that information to the engine's microprocessor control unit.

Metering jet: An orifice within the engine's carburetor that regulates the flow of fuel for combustion. It can be fixed, or of a variable design in which flow is further regulated by a threaded needle at the orifice opening.

Metering rod: A component of the metering jet in some variable-orifice carburetor jet designs. The stepped or tapered rod is raised or lowered by throttle position to change the amount of fuel allowed to flow through the jet orifice.

Methyl alcohol: A toxic, flammable liquid miscible with water and other alcohols, made from the destructive distillation of wood or synthesized from methane. Also known as methanol or wood alcohol.

Methane: A colorless, odorless, flammable gas caused by the anaerobic fermentation of organic materials, and the chief component of natural gas.

Methanol: The common name for methyl alcohol.

Methyl tertiary butyl ether (MTBE): A toxic and volatile liquid used as an additive to gasoline to boost octane and as a substitute for tetraethyl lead. Later used as a component of oxygenated fuels to meet the standards of the Clean Air Act.

Microprocessor control unit (MCU): A microprocessor device used in modern electronically controlled vehicles to evaluate a variety of engine and environmental conditions and moderate the engine's response accordingly. Also called an electronic control unit (ECU) by some manufacturers.

Micro-scale plant: A simple alcohol production facility that uses a single vessel and column for the batch fermentation and distillation of fuel alcohol.

Milling: Reducing a feedstock to smaller particles by mechanically crushing or tearing it.

Miscible: Having the ability, as liquids, to blend uniformly or dissolve into one another.

Molecular sieve: A filtering device that uses zeolite minerals for the adsorption of water, thus allowing ethanol to pass through in a costly and complex alcohol dehydrating process.

Molecule: A group of atoms held together by chemical forces, and the smallest unit of matter that can exist by itself and keep all its chemical properties.

Monosaccharide: A single sugar unit. A carbohydrate which cannot be hydrolyzed, or broken down, to a simpler carbohydrate.

Multi-port fuel injection: A fuel delivery system that injects fuel into each cylinder's intake port rather than at the entry point of the intake manifold.

Neat: Straight or undiluted; alcohol unmixed with water or other fuels.

Needle valve: A control device that uses a tapered needle and orifice to regulate the flow of a liquid or gas.

Neoprene: A synthetic rubber with excellent resistance to chemicals, oils, sunlight and heat, and used in fuel systems that handle alcohol.

Nitrogen oxides: Reactive gases containing oxygen and nitrogen that occur when fuel is combusted at high temperatures. Also called oxides of nitrogen.

Octane: A flammable hydrocarbon and the byproduct of petroleum refining used as a reference fuel to determine comparative octane ratings.

Off-idle circuit: A fuel-air delivery path used in some carburetors to enhance the transition between idle and open-throttle positions.

Oligosaccharide: A carbohydrate containing between two and eight monosaccharide, or single sugar, units.

Open loop mode: The term used when a microprocessor-controlled vehicle is functioning outside the parameters of its oxygen sensor to monitor and regulate fuel mixture and defaults to preset programming.

Oxygen sensor: A device used in an internal combustion engine to measure the concentration of oxygen in the exhaust stream in order to provide real-time information to the vehicle's microprocessor control unit.

Packed column: A distillation column design that uses loosely packed, specially shaped rings or saddles in the chamber to encourage condensation and revaporization of alcohol-water vapors to increase proof strength.

Packing: Plastic, ceramic or stainless steel rings, saddles or other specially shaped objects designed to provide maximum surface area as well as free air space to encourage condensation and revaporization within a distillation column.

Pall ring: A distillation column packing material with a cylindrical shape and perforations, often including internal bridges or foils to increase surface area and turbulence.

Particulate emissions: Miniscule elements of particle matter suspended in a gas, constituting part of exhaust gas emissions.

Pectinase: An enzyme that breaks down the starch-like plant binder pectin into simpler carbohydrates.

Perforated plate column: A distillation column design that uses a series of drilled steel plates or platforms to encourage the vaporization of alcohol and condensation of water at separate stages within the column.

Permaculture: An ecologically based discipline designed to emulate natural patterns to create sustainable human systems.

Phase-change energy: The energy consumed in the transition between the three phases of matter— gaseous, liquid and solid—owing to changes in temperature or pressure.

Phase separation: The parting of a mixture into layers of its individual components.

Phosphoric acid: An acidic chemical compound used to lower the pH level of the mash mixture during fermentation.

pH: A measure of the acidity or alkalinity of a solution based on a scale of 14 points, 7 being neutral. Acids are at the lower end of the scale, alkalines at the higher end.

Pitching solution: A warm, highly active starter solution of wort and yeast used to inoculate a large batch of mash.

Polyolefin plastics: A group of economical thermoplastics such as polypropylene and polyethylene developed for their flexibility, impact resistance, and resistance to chemical damage.

Polysaccharide: A sugar or carbohydrate composed of many monosaccharides.

Positive crankcase ventilation (PCV): A recovery system that collects the noxious vapors that occur in an engine crankcase and recycles them through the intake manifold for combustion.

Positive displacement pump: A fluid pump in which a measured amount of liquid is contained in a chamber, its pressure is increased, and the liquid is expelled through an outlet.

Potentiometer: A device for measuring the force of an electric current, also known as a "pot."

Power valve: A component in an automotive carburetor, governed by vacuum load, that opens when the engine is under a heavy demand to enrich the air-fuel mixture.

Preignition: The premature combustion of the air-fuel mixture within the engine combustion chamber, caused by a source of ignition — often a buildup of deposits — other than the spark plug. The detonation causes an audible pinging and can damage the piston if allowed to continue for an extended period.

Pre-malting: An enzymatic process that breaks down starch in the early phases of hydrolyzation through the addition of an alpha amylase, which keeps the wort thin enough to agitate effectively.

Pressure relief valve: A control device installed on a tank or vessel that opens at a predetermined pressure to allow gas or liquid to escape, and thereby reduce the pressure within the container.

Proof gallon: One gallon of 100 proof alcohol, or a mixture containing 50 percent alcohol by volume, used as a unit of measure for federal tax purposes.

Proof strength: The measure of distilled alcohol content in fuel or beverage alcohol. The proof is twice the percentage level, e.g., 180 proof alcohol contains 90 percent alcohol and 10 percent water by volume.

Protease: An enzyme present in malted grains that digests proteins to aid in fermentation.

Protein: An organic polymer compound composed of alpha amino acids joined by peptide bonds.

Puddling: A condition within the intake manifold of an engine in which the air-fuel vapor mixture condenses and collects as a liquid on manifold surfaces, over-richening the mixture.

Pulse width: The amount of time or duration a fuel injector remains open, sometimes expressed as a percentage.

Quick-disconnect fitting: A removable coupling that employs a lever-and-cam or other positive mechanical means to secure a seal between two pipes or hoses for the sake of convenience, without using welds or threads.

Reboiler: A heat source or heat exchanger built into the bottom of a distillation column for the purpose of creating vapors.

Rectification: The process of vaporizing and condensing an alcohol-water mixture for enrichment, using a countercurrent flow technique.

Rectifier: The portion of a distillation column in which rising vapor is enriched by contact with the countercurrent falling stream of condensed vapor, which returns to the top of the stripper column as a liquid.

Reflux: The condensed alcohol introduced to the top of the distillation column to give the alcohol-water vapors a final condensation to enhance proof strength. Refluxing, in a broader sense, refers to the process of returning condensed vapors to the bottom of the distillation column to be reboiled, vaporized and condensed over again.

Reflux ratio: The ratio of the amount of alcohol condensate being refluxed to the amount being extracted in production; the lower the ratio, the more efficient the distillation.

Refractometer: An instrument used to determine a solution's sugar content by measuring the bend of light through a bead of wort on a refractive index.

Renewable resource: Energy resources that are not depleted by using them, particularly solar, wind or water energy, or biomass energy from trees and plants.

Rhizome: A buried horizontal plant stem that has buds, nodes and leaves and grows both roots and shoots.

Rich mixture: A fuel-air mixture in an internal combustion engine that contains less air or more fuel than the stoichiometric ideal.

Riser: In a plate or tray distillation column, a short conduit designed to pass vapors from one plate to the plate directly above it, where a bubble cap assembly forces the vapors back down through a liquid alcohol-water mixture to promote fractionation.

Run: A single distillation from beginning to end, or the time it takes to complete the distillation process in a batch operation.

Saccharification: The conversion step in the process of fermentation in which glucoamylase enzymes reduce complex sugars to simple fermentable sugars.

Saccharometer: An instrument that measures the amount of sugar in a solution by determining its specific gravity.

Saccharomyces cerevisiae: The common yeast used for fermentation and breadmaking.

Saddle packing: A type of distillation column packing material formed in a semicircular shape, convex on one surface and concave on the other, often with serrated edges to enhance turbulence.

Screw press: A mechanism used to force water from feedstock pulp or residue, with pressure applied by means of a heavy threaded shaft.

Separator: A rotary machine that uses centrifugal force to separate liquids from solids in preparing mash.

Silage: Green or mature fodder compressed and fermented, then stored in silos for use as winter livestock feed.

Simple sugars: Fermentable sugars, or monosaccharides such as glucose that can be converted to alcohol without further reduction.

Simultaneous saccharifcation and fermentation: A process arising from the development of improved enzymes and yeasts that allows fermentation to occur immediately after liquefaction, thus eliminating the saccharification step.

Slurry: A mixture of liquid and suspended solids, as in traditional grain mash.

Small producer tax credit: A tax credit most recently modified with the Energy Policy Act of 2005 that allows an eligible Small Ethanol Producer (production of less than 60 million gallons per year) a federal income tax credit equal to $ .10 per gallon for the first 15 million gallons produced.

Sparge: To bubble or distribute air or gas into a liquid for the purpose of agitation and contact.

Spark-ignition engine: A conventional engine, fueled by gasoline, liquefied petroleum gas, or other fuel, in which the air-fuel mixture is ignited in the combustion chamber by means of a spark plug.

Spark plug: An electrical device in an automotive engine that delivers a high-voltage spark directly to the combustion chamber to ignite the air-fuel charge.

Specific gravity: The ratio of the density of a material to the density of distilled water at 60°F; also known as relative density.

Spent grains: The nonfermentable solid material left over after the fermentation of a grain mash.

Starch: A polymerized glucose or polysaccharide that occurs as structural granules and a carbohydrate storage medium in many plant cells.

Still: A derivative of the word distillery, used to describe the distillation column and any associated mash containers and condensers.

Stillage: The residual materials left after distillation, containing distiller's solubles and fines, and water from the mash. Also called whole stillage.

Stoichiometric ratio: A chemically balanced air-fuel ratio in the combustion chamber of an engine. With gasoline fuel, the ratio is 14.7:1. With ethanol, it is 9.0:1.

Stover: The residue, such as stalks and leaves, left over after harvesting corn, grains, sorghum and other seed crops, usable as a biomass fuel when dried.

Stripper: The portion of a distillation column which removes vapors from the descending liquids, where they rise into the rectifier section.

Substrate: The substance on which an enzyme acts, derived from the feedstock. Sugar is the substrate for the zymase enzymes in yeast, and starch is the substrate for amylase enzymes.

Sucrase: An enzyme which breaks down the disaccharide sucrose to monosaccharides.

Sucrose: The most common disaccharide, produced from sugar cane, sugar beets and sorghum. The enzyme sucrase, produced by yeast, breaks down sucrose to one molecule each of glucose and fructose for fermentation.

Sulfuric acid: An acidic and corrosive chemical compound used to adjust pH levels when cooking and fermenting mash.

Supercharger: A belt- or gear-driven compressor that forces air under pressure into the intake manifold of an engine for improved aspiration.

Tetraethyl lead (TEL): A highly toxic lead compound used to increase gasoline's anti-knock qualities.

Thermocouple: A device comprised of two dissimilar metals used to determine the temperature of a gas, liquid or solid by reading thermoelectric voltage output.

Thermophilic: Able to thrive at high temperatures, as bacteria and other organisms that are thermophiles.

Throttle-body fuel injection (TBI): An electronic fuel injection system that delivers fuel under pressure to the throttle port at the entry to the intake manifold.

Throttle plate: A butterfly valve located at the base of a carburetor or beneath a throttle body injector that regulates the amount of air-fuel mixture entering the intake manifold.

Top dead center (TDC): The highest point of the piston travel in an internal combustion engine, when the crank is at the top of its stroke.

Turbocharger: A compressor driven by an engine's exhaust gases that forces air into the intake manifold for enhanced aspiration.

Turbulent flow: The tumultuous passage of an alcohol-water mixture over packing or perforated plates in a distillation column, constantly changing liquid velocity in magnitude and direction; necessary to break laminar flow into a turbulence to give alcohol molecules an opportunity to come to the surface of the liquid stream and escape as vapor.

Vacuum advance: A mechanical device operated by engine vacuum that advances ignition timing with an increase in engine rpm.

Vacuum breaker: A safety device used to relieve vacuum pressure in a tank or plumbing line to prevent unwanted backflow or structural collapse.

Vacuum distillation: An alcohol distillation technique that uses reduced vapor pressure to lower the boiling points of water and alcohol for the purpose of conserving thermal energy.

Vaporize: To change from a solid or liquid to a vapor through the application of heat or by reducing pressure. Also called volatilization.

Vapor lock: The interruption of fuel flow in an internal combustion engine caused by the formation of vapor or gas bubbles that block the fuel line.

Vapor pressure: A measure of the evaporative qualities of a combustible or fuel, or the pressure of vapor in equilibrium with its liquid or solid state. When the vapor pressure of a liquid equals that of its surroundings, it will boil.

Venturi: A specially shaped narrowed passageway within a carburetor throat that cause airflow to increase in velocity and air pressure to drop, pushing fuel out of the float reservoir with atmospheric pressure.

Viscosity: A measure of resistance to flow in a fluid, due to internal friction.

Volatility: A measure of evaporation tendencies in a substance, intensified by a low boiling point or a high vapor pressure.

Waste gate: A relief valve in an automotive turbocharger that regulates the pressure exerted by the engine exhaust gases on the turbine, thereby controlling the boost pressure to the manifold.

Watt: A unit of power, equal to one joule per second, in the meter-kilogram-second system of units.

Weir dam: A short wall or barrier on the plate of a continuous distillation column that regulates the depth of liquid on the plate and allows solids in the mash to pass over.

Wet distiller's grains: The solid materials collected after grain mash has been distilled and filtered, prior to dewatering or drying.

Wet-milling: A process used mainly in industrial-scale alcohol production, in which feedstock grains are separated into individual co-products prior to fermentation and only the starch is used in the actual fermentation process.

Whey: The liquid portion of milk, separated from the casein or coagulated curd in cheesemaking, and useful for its lactose and residual protein content.

Wide-open throttle: The condition of maximum throttle opening in an engine, when the greatest amount of airflow occurs.

Wine gallon: A unit of measure for fuel and beverage alcohol production equivalent to one US gallon of liquid.

Wort: A mixture of grain and water, prepared by grinding and heating the feedstock, prior to adding yeast. With the addition of yeast it is called mash, and after fermentation it is a beer.

Xylose: A five-carbon sugar extracted from wood or straw grasses once the cellulosic material has been broken down by special enzymes.

Yeast: Small unicellular fungi that reproduce by budding and, through metabolism and enzymatic action, convert sugar into alcohol and carbon dioxide in the process called fermentation.

Yeast starter: An introductory dose of healthy yeast solution used to create vigorous fermentation and reduce fermentation time to a minimum.

Zymase: An enzyme complex that converts glucose sugar to alcohol and carbon dioxide.

Zeolite: A naturally occurring hydrous mineral characterized by an aluminosilicate tetrahedral framework and loosely bonded water molecules allowing reversible dehydration; used to trap water in molecular sieves for alcohol dehydration in the production of anhydrous alcohol.

Bibliography

Books

Aubrecht, Gordon J., *Energy: Physical, Environmental, and Social Impact,* Third Edition, Pearson Education Inc., 2006. A broad scientific overview of global energy trends and resource supplies. Examines production, global consumption, the impact of politics and economics, and long-term prospects.

Bernton, Hal, William Kovarik, and Scott Sklar, *Forbidden Fuel: Power Alcohol in the Twentieth Century,* Caroline House Publications, 1982. An eye-opening and thoroughly researched work on the political, economic and agricultural impact of alcohol motor fuels in the Twentieth Century. It is a tale of corporate deceit and international intrigue at some levels, and history buffs will appreciate the scale of the authors' treatment of the subject.

Blume, David, *Alcohol Can Be a Gas,* International Institute for Ecological Agriculture, 2007. This somewhat self-promoting work is a complete but disorganized compendium on politics, economics, society and ethanol fuel. It was assembled over a period of time using a variety of resources, though some have little to do with the subject of alcohol fuel. Contains information on feedstocks, fermentation, distillation, application, and many case studies, but it could have been halved in size (and cost) without losing anything of value. At risk of being a combination "what if" wishbook and political screed, save for the details on alcohol fuel.

Brown, Michael H., *Brown's Alcohol Motor Fuel Cookbook,* Desert Publications, 1979. One of the earliest sources of information on automotive conversions for ethanol fuel, but not a particularly exceptional reference, as it is rife with generalizations. Does include some information on fermentation and distilling, but with very unsophisticated apparatus.

Doxon, Lynn Ellen, *The Alcohol Fuel Handbook,* 2nd ed., Tallgrass Research Center, 1980. A working guide to making alcohol fuel that is better than most of its contemporary references, but still lacks critical information on engine conversions, distillation equipment details, and fermentation procedures.

Gingery, Vincent R. *The Secrets of Building an Alcohol Producing Still,* David J. Gingery, 1994. A good basic overview of fermentation and distillation, including a building plan for a simple still. Contains entry-level but nonetheless useful information.

Goettemoeller, Jeffrey and Adrian Goettemoeller, *Sustainable Ethanol,* Prairie Oak Publishing, 2007. A timely and well-researched study of ethanol fuel and how it affects US agricultural and economic status. Contains a lot of information on energy policy, sustainable farming practices, and advances in distillation technology including cellulosic ethanol.

Goosen, Clarence D., *Goosen's EtOH Fuel Book,* The Harvester Press, 1980. A self-published volume by a former staffer at *The Mother Earth News* magazine who was responsible for much of that publication's work in the fuel alcohol field. This initial volume covers enzymes, basic feedstocks, grain malting, distillation, permitting and designing and operating a packed column still. Some of the information is dated, but it is thoroughly researched and the book contains a good glossary.

Goosen, Clarence D., *Goosen's EtOH Fuel Book,* Issue 2, The Harvester Press, 1980. This second self-published volume covers yeasts, detailed distillation columns, and designs for a mash cooker and a small vacuum still. The author approaches the subject in an honest, technical manner and organizes the book as a series of notes. Regrettably, this has the effect of creating some degree of disorganization among the three volumes, as they are not well coordinated.

Goosen, Clarence D., *Goosen's EtOH Fuel Book,* Issue 3, The Harvester Press, 1981. This third volume includes some useful tables for hydrometer temperature correction and vacuum distillation, a starch cooking recipe, testing procedures, details on a continuous vacuum still, more distillation theory, and some basic automotive conversion information.

Lowther, Granville, *The Encyclopedia of Practical Horticulture,* Encyclopedia of Horticulture Corporation, 1914. Contains an interesting section on the use of farm and garden crops for the production of commercial alcohol for fuel and industry.

Nixon, Michael and Michael McCaw, *The Compleat Distiller,* Amphora Society, 2001. A concise and surprisingly thorough summary of small-scale fermentation and distillation. Directed toward home distillers, it is nonetheless valuable for its clear explanations of processes and small-scale equipment.

Rombauer, Irma S., Marion Rombauer Becker, and Ethan Becker, *The Joy of Cooking,* Anniversary Edition, Scribner, 2006. The timeless classic, recognized worldwide. Modern editions include a comprehensive listing of the nutritional value of a large sampling of grains, fruits and vegetables.

Stone, Michael and Sally Stone, *The Essential Root Vegetable Cookbook,* Clarkson Potter

Publishers, 1991. An appealing cookbook in its own right, but also valuable for the data on nutritional value of a wide variety of starch crops.

Toboldt, William K., *Automotive Encyclopedia*, Goodheart-Willcox, 2005. An excellent reference for detailed technical information on all aspects of automobile engine theory, operation and components.

United States Government, *2008 Cellulosic Ethanol* (CD), Progressive Management, 2007. A CD that is essentially a collection of technical and policy papers on cellulose-based ethanol production, from viewpoints that are economic, agricultural, industrial and energy-related.

US Dept. of Energy, *Internal Combustion Engines for Alcohol Motor Fuels*, US Dept. of Energy, Office of Alcohol Fuels, Nov. 1980. This is a technical report on the viability of alcohol fuels, with particular reference to the NAHBE engine optimized for ethanol fuel use. It is a general compilation of various technical papers of the period.

Periodicals

Acres, USA, P.O. Box 91299, Austin, Texas 78709. Tel. (800) 355-5313. Website: acresusa.com. A long-standing monthly publication of eco-agriculture covering soil management, composting, insect control, crop advice, value-added processing. Excellent information for the small sustainable grower.

BackHome Magazine, P.O. Box 70, Hendersonville, NC 28793. Tel. (800) 992-2546. Website: BackHomeMagazine.com. Touted as a hands-on guide to sustainable living, this bi-monthly was founded some years ago by some of the original *Mother Earth News* staff. Includes a lot of good how-to information on renewable energy, alternative and biomass fuels, organic growing, green building, and small farm operations.

Countryside & Small Stock Journal, 145 Industrial Dr., Medford, WI 54451. Tel. (715) 785-7979. Website: countrysidemag.com. A journal of homesteading and the farmstead lifestyle, with articles on small-scale agriculture, livestock, and some renewable energy coverage, mostly in a rural setting.

Ethanol Producer Magazine, 308 2nd Avenue North, Suite 304, Grand Forks, ND 58203. Tel. (701)746-8385. Website: ethanolproducer.com. A trade magazine that covers the entire gamut of the ethanol industry, including policy, research, markets, funding, new technologies, and current production. Not necessarily aimed at the small producer but offers good insight into the future of ethanol fuel.

Ethanol Today Magazine, 5000 S. Broadband Lane, Suite 224, Sioux Falls, SD 57108. Tel. (605) 334-3381. Website: ethanoltoday.com. An online publication of the American Coalition for Ethanol, advocacy group for ethanol fuel use. Good source for fuel producers, blenders and potential ethanol marketers, with excellent resource material on federal and individual state policy and vehicle applications.

Farm Show Magazine, Johnson Bldg., P.O. Box 1029, Lakeville, MN 55044. Tel. (800) 834-9665. Website: farmshow.com. An interesting and sometimes offbeat publication of do-it-yourself projects and product ventures in rural agricultural communities across the nation. A good source for ideas and a place to see clever adaptations of farm equipment to repurposed uses.

Home Power, P.O. Box 520, Ashland, OR 97520. Tel. (800) 707-6585. Website: homepower.com. An excellent bi-monthly publication that deals mainly with renewable energy (solar, wind, microhydro) and its technology, including occasional biomass articles.

Mother Earth News, 1503 SW 42nd St., Topeka, KS 66609. Tel. (785) 274-4300 Website: motherearthnews.com. The original back-to-the-land magazine, with a solid editorial history of ecological and sustainable living. Covers organic food and farming, some renewable energy, and green living.

Endnotes

Chapter One: About Alcohol Fuel

1. Bush, George W., washingtonpost.com, (online) "State of the Union 2007" address. (cited April 15, 2009) , January 23, 2007.
2. cato.org, Cato Institute, AEI Journal on Government and Society, "Perspectives on Current Developments," p. 12, December 1981.
3. bipartisanpolicy.org, Clean Air Act of 1990, Reformulated Gasoline Standards (access February 19, 2008).
4. Kovarik, William:, *Fuel of the Future* note 42, Rufus Frost Herrick, Denatured or Industrial Alcohol, (New York: John Wiley & Sons, 1907), p. 307.
5. Kovarik, William:, *Fuel of the Future* note 48, Col. Sir Frederic Nathan, "Alcohol for Power Purposes," The Transactions of the World Power Congress, London, Sept. 24 - Oct. 6, 1928.
6. Kovarik, William:, *Fuel of the Future*, note 1, "Ford Predicts Fuel from Vegetation," New York Times, Sept. 20, 1925, p. 24.
7. Kovarik, William:, *Fuel of the Future*, note 73, H.B. Dixon, "Researches on Alcohol as an Engine Fuel," SAE Journal, Dec. 1920, p. 521.
8. Kovarik, William: , *Fuel of the Future*, note 82, M.C. Whitaker, "Alcohol for Power," Chemists Club, New York, Sept. 30, 1925. Cited in Hixon, "Use of Alcohol in Motor Fuels: Progress Report No. 6," Iowa State College, May 1, 1933.
9. Lewis, Jack, EPA Journal, Jack Lewis, May 1985, US Environmental Protection Agency.
10. Lewis, Jack, EPA Journal, Jack Lewis, May 1985, US Environmental Protection Agency.
11. eia.doe.gov, ref 95-618, access February 18, 2008.
12. Ethanol Summit, May 2008 — domesticfuel.com.
13. Ethanol Summit, May 2008 — domesticfuel.com.
14. Reuters, July 6, 2008 "Brazil Local Demand to Drive Ethanol Production."
15. "Anúario Estatístico 2008: Tabelas 2.1-2.2-2.3 Produção por combustível - 1957/2007" (PDF). ANFAVEA - Associação Nacional dos Fabricantes de Veículos Automotores (Brasil). (Access December 19, 2008).
16. "Biofuels Report," 2008, p. 71, TheWorld Bank.
17. USDA SARE Report, "Livestock and Feedstock: Distillers Grains and Fuel Ethanol."

Chapter Three: Federal and State Requirements

1. Internal Revenue Code, Subchapter B, Chapter 51.
2. United States Code, 26 U.S.C. Section 5181, Ref. Public Law 96-223.

Chapter Four: Do-It-Yourself Economics

1. *Fuel From Farms*, US Department of Energy, SERI, February 1980, p. 8.
2. Ibid. p. 11.

Chapter Five: Feedstocks and Raw Materials

1. USDA "The Economic Feasibility of Ethanol Production from Sugar in the United States," July 2006.

2. USDA Publication 327, 1938.
3. Anderson, I.C., *Ethanol From Sweet Sorghum,* Iowa State Univ.
4. Encyclopedia Britannica, Molasses .
5. USDA Pub. 327, 1938.
6. USDA *Motor Fuels from Farm Products* Misc. Pub. No. 327, December 1938.
7. Fairbank, W.C., *Jerusalem Artichoke for Ethanol Production,* Univ. of California Cooperative Extension, Vegetable Research and Information Center. 1982.
8. Jones, Hamilton, *Dukes,* "Sweet Potato Cultivars for Ethanol Production," *Proceedings of the Third Annual Solar Biomass Workshop* Atlanta, April 1983, p. 196.
9. "Ethanol Feedstock from Citrus Peel Waste," USDA Agricultural Research Service, May 19, 2008.

Chapter Six: Starch, Sugar and Fermentation

1: *Goosen's EtOH Fuel Book,* vol. 2, "Alcohol Production," p.13
2. *Goosen's EtOH Fuel Book,* vol. 2, "Microbiology," p.5
3. *Goosen's EtOH Fuel Book,* Vol. 2, "Microbiology," p.7

Chapter Seven: Distillation

1. *Goosen's EtOH Fuel Book,* vol. 2, p. A2.
2. Glover, Thom. J., *Pocket Ref,* "Firewood/Fuel Comparisons," p. 191.

Chapter Nine: Alcohol as an Engine Fuel

1. US Department of Energy, Office of Alcohol Fuels, "Descriptive Summary of Internal Combustion Engines," Fig. 6 *Internal Combustion Engines for Alcohol Motor Fuels.*
2: Compton, Timothy, "Flight Performance Testing of Ethanol/Avgas Fuel Blends During Cruise Flight." p. 45, Baylor University, June 5, 2008.
3: Stoyke, Godo, *The Carbon Buster's Home Energy Handbook,* p. 22.
4: Jeuland, N., X. Montagne and X. Gautrot, Institut français du pétrole, *Potentiality of Ethanol as a Fuel for Dedicated Engine* from the conference proceedings "Which Fuels For Low-CO2 Engines?" Oil & Gas Science and Technology – Rev. IFPVol. 59 (2004), No. 6, pp. 559-570.
5. Ibid.
6. *Internal Combustion Engines for Alcohol Motor Fuels,* Nov. 1980, US Dept. of Energy, Office of Alcohol Fuels pp. 3-107.
7. Compton, Timothy, "Flight Performance Testing of Ethanol/Avgas Fuel Blends During Cruise Flight." p. 7, Baylor University, June 5, 2008.
8. American Petroleum Institute (API) www.api.org/Newsroom; (access June 5, 2008).

Chapter Ten: Case Studies

1. cleanhouston.org/energyfeatures/ethanol2.htm (access April 8, 2008).
2. Business Week online, Dec. 18, 2006 (access April 8, 2008). April 8, 2008.
3. Wackett, Lawrence P., and Douglas C. Hershberger, ***Biocatalysis and Biodegradation,*** Lawrence P. Wackett and C. Douglas Hershberger, ASM Press, p. 41.
4. cleanhouston.org/energy/features ethanol2.htm, (access April 4, 2008) April 2008.
5. biomass.novozymes.com/files/documents/Thematic%20paper%20-%20Biofuel%20technology%20development.pdf (access April 10, 2008). April 10, 2008.
6. Purdue News, "Purdue yeast makes ethanol from agricultural waste more effectively," June 28, 2004.
7. iogen.ca/cellulose_ethanol/what_is.
8. Carey, John and Adam Aston, "Put a Termite in your Tank," *Business Week.* December 18, 2006.
9. "Basic Petroleum Statistics," *Finished Motor Gasoline,* 2007 Energy Information Administration, Department of Energy, US Government.
10. Iogen.ca/key_messages/overview/m4_fuels_vehicles.html.
11. Smith, Jeffrey M., *Genetic Roulette* (Yes! Books, 2007), p. 85.

Index

accelerator pumps, 175–176
adenosine triphosphate, 45, 71
aerobic conditions, 71, 75, 225
Alcohol and Tobacco Tax and Trade Bureau (TTB), 28, 34, 89, 154, 217–219. *See also* Alcohol Fuel Producer (AFP) permit
Alcohol Beverage Control Commission, 36, 37
alcohol-fired distillation, 38, 115, 126
Alcohol Fuel Producer (AFP) permit, 25–26, 28–36, 119–120
  denaturing process, 34–35
  environmental impact, 38
  fire codes and zoning, 37–38
  forms and reports, 29–33, 35
  insurance and bonding, 35
  levels of, 28–29, 119–120
  plant premises and storage, 33–34
  record keeping, 35
  state and local jurisdiction, 37–38
alcohol storage, 33–34, 38, 108, 127, 139, 153, 153–154, 164, 208
alcohol yield. *See also specific crops*
  calculating, 100
  testing for, 89–90
alfalfa, 57
alpha amylase, 47, 55, 64, 66, 67, 77, 78, 79, 81, 85, 91, 92–93, 94, 95, 214, 225, 226, 233, 235
aluminum-silicon, 102, 108
Alvarez, B.V., 6
ammonium sulfate fertilizer, 72, 99
amyloglucosidase (glucoamylase), 47, 67–68, 77, 78–79, 81, 82, 85, 91, 92, 93, 94, 95, 207, 225, 228, 231, 236
amylopectin, 225
amylose, 65, 225
anaerobic conditions, 45, 70, 71, 82, 93, 100, 207, 225, 233
anhydrous alcohol, 10, 15, 16, 104, 111, 164, 165, 204, 228, 239
anti-knock properties, 8, 10, 14, 158, 202
apples, 40, 46, 48, 52, 53, 66, 83
apricots, 46, 52, 53
auger presses, 96, 122
autoignition temperature, 35, 157, 158, 225
azeotropic ethanol, 15–16, 89, 102, 109–112, 225, 226
bacterial contamination, 72–73
bagasse, 49, 226

barley, 40, 46, 54, 55, 66, 79–82, 226, 233
batch distillation, 23, 30, 41, 43, 107, 108, 109, 123, 135, 136, 143, 146–149, 150, 153, 226, 233, 234, 236
  vs. continuous, 24–25
  finishing off and low wines, 148–149
  initiating and controlling heat, 147
  maintaining column equilibrium, 147–148
  preparing the tank, 147
beans, 66, 83
beer, 38, 45, 72, 93, 101, 107, 147, 148, 206, 207, 208, 212, 226, 233, 239
before top dead center (BTDC), 168, 189
benzene, 10, 14, 102, 225, 226
Bernoulli Theorem, 171
beta amylase, 67, 68, 77, 78, 79, 81, 226, 233
beverage, bakery and candy waste, 60–61
Biocon, 68, 91–92
biodiesel fuel, 21, 22, 204
Blaser, Richard F., 161
boiling points, 101–102, 103, 226
bottom dead center (BDC), 167, 226
Brazil, 15–18, 49, 62, 109, 115, 165, 201, 202, 203
British Thermal Unit (BTU), 157, 226
Brix scale, 89, 90, 226
Bryan, Charles W., 12
bubble cap columns, 105, 226, 236
buckwheat, 46
Bureau of Alcohol, Tobacco and Firearms (BATF), 28, 29–30, 37, 226
Burkholder, Dennis, 5
Bush, George W., 3, 19
by-product use, 116–118, 209
Canada, 37, 49, 210–214
carbon dioxide, 25, 38, 43, 45, 47, 48, 55, 62, 69, 70, 71, 72, 75, 93, 100, 102, 112, 117–118, 160, 203, 208, 209, 216, 227, 229
carbon monoxide, 4, 13, 14, 161–162, 227
carburetion, 168–184, 227
  accelerator pump, 175–176
  choke, 172, 179–180
  dual-fuel carburetors, 183
  electronic feedback carburetors, 182–184
  float level adjustment, 179, 183
  fuel preheating, 180–182
  idle circuit, 168, 176, 177–179, 183, 198, 232
  jet diameter vs. jet area, 171
  main circuit, 168–175
  power valve, 168, 171, 175, 176–177, 235
  thermostat replacement, 180
carrots, 46
cassava, 55, 56–57
catalytic converters, 162, 227
cattails, 58–59
caustic lime (calcium oxide), 88, 99, 104, 111
cellulosic ethanol, 57, 210–214, 227
centrifugal compressors, 196–197
centrifugal pumps, 75, 95, 127, 130, 227
Chepeka, Mandy, 214
Chevrolet, 8, 161, 182, 183, 192
chlorine, 69–70, 73, 76, 86, 87, 92, 104, 149, 207
chokes, 172, 179–180
citrus and tropical fruit, 40, 46, 60
Clean Air Act, 4, 13
Clean Water Act, 33, 38
co-generation, 116, 125, 227
Code of Federal Regulations (CFR), 29, 34, 35
cold-starting, 15, 17, 163–164, 165, 179, 192–194, 198, 199, 200, 231
combustion chamber, 158, 159, 160, 161, 162, 167, 169, 175, 186, 189, 191, 194, 203, 216, 227, 230, 232, 235, 237
combustion properties, 155–162
  emissions, 160–162

heat of combustion, 156–158
latent heat of vaporization, 159–160
octane rating, 158–159
volatility, 159
common rails, 184, 187, 227
complex sugars, 45, 66, 227, 228, 236
compression-ignition engines, 166, 184, 194, 195–197, 202–204, 227, 229, 231
compression ratio, 158, 159, 161, 190, 194–197, 198, 202, 227
condensers, 138–141, 142, 181, 228
continuous distillation, 24–25, 28, 107–109, 228
control coils, 128, 228
conversion. *See* saccharification
cooker vats, 113, 128–132, 228, 233
cooling coils, 53, 68, 72, 93, 95, 98, 110, 228
corn, 28, 55, 55, 65, 66, 83, 90–91, 117
conventional corn mash recipe, 92–94
economy of, 2, 4, 11, 12, 13, 18–19
ethanol yield, 40, 46, 54
no cooking corn mash recipe, 94
Painterland Farms, 206–209
corrosion, 164–165
crabapples, 46, 66, 83
cranberries, 46
Crombie, Lance, 4–5, 114
crop residue, 58, 116, 237
Crude Oil Windfall Profit Tax Act, 28
crushers, 96
Daschle, Tom, 4
dates, 46, 52, 83
degradation and corrosion, 164–165
dehydration, 102, 208, 228, 239
denaturing, 28, 29, 34–35, 37, 38, 146, 153, 162, 208, 228
Department of Energy, 7, 161
dephlegmators, 141–143, 149
dewatering, 122, 147, 228, 239
dextrins, 63, 66, 67, 71, 77, 78, 81, 85, 92, 93, 225, 226, 228, 231, 232
diaphragm pumps, 127, 130, 165, 228
diesel, 116, 155, 227
biodiesel fuel, 21, 22, 204
block conversion, 195–197
engines, 166, 184, 194, 195–197, 202–204, 227, 229, 231
diffusion, 49, 51, 96–97, 228
digestible energy, 18
direct refluxing, 127, 144, 228
disaccharides, 63, 87–89, 228
lactose, 51, 59, 60, 63, 232, 239
maltose, 63, 72, 77, 78, 79, 85, 226, 228, 233
sucrose, 63, 67, 79, 232, 238
distillation, 23, 24, 25, 30, 41, 43, 101–118, 123, 135, 136, 143, 146–149, 150, 153, 226, 228, 233, 234, 236
azeotropic, 109–112
batch. *See* batch distillation
by-product use, 116–118
columns. *See* distillation columns
continuous, 24–25, 28, 107–109, 228
heat sources, 112–116
vacuum, 149–153
distillation columns, 132–143
bubble cap, 105, 226, 236
condensers and heat exchanges, 138–141
dephlegmators, 141–143
packed. *See* packing
perforated plate, 30, 105, 107, 136–137, 151, 208, 235, 238
stripper, 108, 127, 135, 141, 143, 148, 149, 237
distiller feeds, 116–117
Condensed Distiller's Solubles (CDS), 116, 228
Distiller's Dried Grains (DDG), 25, 116, 229

Distiller's Dried Solubles (DDS), 229
Distiller's Solubles (DS), 116, 229
Dried Distiller's Grains with Solubles (DDGS), 18, 52, 54, 116, 229
Wet Distiller's Grains (WDG), 116–117
distributors, 183, 184, 189–190, 229
Dole, Bob, 4
downcomers, 104, 136, 137, 229
Doyle, Ned, 176
drying ethanol, 104, 111–112
dual-fuel carburetors, 183
economics of ethanol, 18, 39–44, 215
business management, 209–210
buying vs. building equipment, 41–42
cost per gallon, 42–44
sourcing raw materials, 41
tax credits, 4, 44, 200, 215–216, 237
Electronic Control Units (ECU), 182, 200, 202, 203, 229, 233
electronic feedback carburetors, 182–184
electronic fuel injection (EFI), 184–189, 229, 230, 238
Embden, Georg, 69
Embden-Meyerhof Pathway, 47, 69, 79, 229
emissions, 160–162
energy balance, 17, 18, 215, 229
Energy Policy Act, 4, 14, 44
Energy Returned On Energy Invested (EROEI), 62, 215
Energy Tax Act, 13
engine modifications, 165–199
carburetion, 168–184
cold-starting systems, 192–194
compression ratio, 194–197
diesel engines, 195–197, 202–204
hot air induction, 191–192
ignition timing, 189–191
small and single-cylinder engines, 197–199
enrichment, 110, 137, 138, 229, 236
environmental considerations, 216
Environmental Protection Agency (EPA), 13–14, 24, 161, 164, 216
enzymes, 25, 45, 46, 52, 63, 76–82, 230
adenosine triphosphate, 45, 71
alpha amylase, 47, 55, 64, 66, 67, 77, 78, 79, 81, 85, 91, 92–93, 94, 95, 214, 225, 226, 233, 235
beta amylase, 67, 68, 77, 78, 79, 81, 226, 233
cellulose-exploiting, 210–214, 227
diatase, 228
effects of environment on, 76–78
Genetically Modified Organisms (GMO), 211, 214, 231
glucoamylase (amyloglucosidase), 47, 67–68, 77, 78, 78–79, 81, 82, 85, 91, 92, 93, 94, 95, 207, 225, 228, 231, 236
inulinase, 51
invertase, 232
maltase, 79, 233
pectinase, 235
protease, 72, 236
sucrase, 79, 238
zymase, 238, 239
equilibrium, 108, 110, 147–148, 230
equipment, distillery, 119–154
accessories, 127–128
augers, 120
batch-run stills, 146–149
buying vs. building, 41–42
cleanup and maintenance, 149
columns. *See* distillation columns
heat sources, 125–126
pumps. *See* pumps
reboilers, 143–144
refluxing, 144
storage. *See* storage

tanks. *See* tanks
vacuum distillation, 149–153
ethanol blends
E-85, 13–15, 103, 163, 192, 200, 229
E-95, 203, 229
gasohol (<20% alcohol), 3, 7, 13–15, 17, 104, 164, 209, 231
ethyl tertiary butyl ether (ETBE), 13
exhaust gas recirculation (EGR), 230
exhaust gas temperature (EGT), 160, 161, 162, 171, 184, 187, 230
federal and state requirements, 27–38
feedback carburetors, 182–184, 192, 230
feedstocks and raw materials, 40, 45–62, 66, 83, 96–97, 122, 206–209, 230
fermentation, 45–46, 69–70, 71, 230
basic sugar technique, 96–100
carbon dioxide, 117–118
conventional corn mash recipe, 92–94
no cooking corn mash recipe, 94
potato mash recipe, 94–96
saccharification, 67–68, 74, 82, 85, 92, 93, 94, 207, 228, 236, 237
starch crops, 91–96
testing procedures, 82–90
zymolysis, 93, 214
fermentation locks, 70, 100, 118, 230
fermentation tanks, 73, 75, 94, 208, 230
Fiat, 17, 201
figs, 46, 52, 83
filtering, 85, 89, 99, 107, 112, 150, 154, 164, 186
fire codes, 37–38
fireboxes, 23, 112, 113, 124, 129, 132, 143, 149, 230
flash points, 159, 163, 230
Flexible Fuel vehicles (FFV), 14–16, 17, 109, 163, 199–202, 213, 230
flinty starch, 64, 230
float levels, 179, 183
flow control valves, 128
fluoroelastometers, 165
fodder beets, 52
forage crops, 57–58
Ford Motors, 4, 9, 11, 15, 17, 169, 182, 189, 192, 195, 199–200
four-stroke engines, 197–198
fractionation, 102, 108, 110, 143, 144, 152, 229, 230, 236
fructose, 47, 51, 55, 60, 67, 69, 71, 79, 231, 232, 238
fuel economy, 162–165
fuel injection, 4, 14, 162, 163, 165, 168, 181, 192, 193, 194, 200, 227, 231, 236
electronic (EFI), 184–189, 229, 230, 238
multi-port, 185, 196, 234
throttle-body (TBI), 184–185, 238
fuel preheating, 180–182
galactose, 63, 71
gasohol, 3, 7, 13–15, 17, 104, 164, 209, 231
Gasolase (glucoamylase), 91, 93, 94
gasoline, 3–4
combustion properties, 155–162
leaded, 10–11, 12
performance and fuel economy, 162–163
unleaded, 13, 14, 28, 34, 35, 104, 158, 164, 183, 194, 200, 208, 230, 231
gate valves, 126, 231
gelatinization, 65–66, 67, 77, 92, 93, 94, 95, 130, 231
Genencor International Inc., 94, 207
General Motors (GM), 10, 16, 17, 183, 195
Genetically Modified Organisms (GMO), 211, 214, 231
Germany, 4, 9, 213
germination, 79, 81, 226, 231
glow plugs, 194, 196, 231
glucoamylase, 47, 67–68, 77, 78–79, 81, 82, 85, 91, 92, 93, 94, 95, 207, 228, 231, 236

glucose, 47, 51, 55, 60, 63, 65, 66, 69, 71, 72, 79, 85, 90, 92, 93, 212, 225, 227, 228, 229, 231, 232, 233, 237, 238, 239
Goosen, Clarence, 5, 114, 135
grain (milo) sorghum, 46, 50, 54, 55, 120
grains, 54–55. *See also* specific grains
grapefruit, 66, 83
grapes, 40, 46, 52, 83
grinding, 41, 42, 48, 52, 56, 64–65, 81, 92, 95, 121–122, 206, 207, 231
Hamilton, Alexander, 27
hammermills, 48, 64–65, 92, 97, 121, 130, 146, 231
heat exchangers, 73, 95, 98, 107, 108, 109, 110, 112, 114, 116, 125, 127, 128, 135, 139–141, 144, 147, 148, 181, 204, 228, 231
heat of combustion, 156–158, 231
heat sources, 112–116, 125–126
  agricultural residue, 113, 115–116
  alcohol, 38, 115, 126
  co-generation, 116, 125, 227
  coal, 113
  geothermal, 113
  heat-transfer fluid, 114, 123–124, 125–126
  hydroelectricity, 116, 125
  municipal solid waste, 113
  natural gas, 13, 38, 47, 112, 118, 155, 156, 201, 204, 233
  petroleum, 126
  solar, 113, 114–115, 125
  steam, 115, 126
  wind, 113, 115, 116, 125
  wood, 38, 112–114, 123, 125
high energy ignition (HEI) systems, 190
history of ethanol fuel, 4–13
hot air induction, 191–192
hydration, 94
hydrolysis, 52, 57, 64, 76, 77, 78, 92–93, 130, 206, 211, 212, 231, 232
hydrometers, 35, 89, 90, 217–219, 226, 232
hydrous ethanol, 15–16, 89, 102, 109–112, 225, 226
idle circuits, 168, 176, 177–179, 183, 198, 232
ignition timing, 159, 161, 165, 168, 182, 188, 189–191, 198, 200, 201, 232, 238
insurance and bonding, 35
internal combustion engines (ICE), 10, 40, 116, 158, 226, 227, 230, 232, 233, 234, 236, 238
  diesel, 166, 184, 194, 195–197, 202–204, 227, 229, 231
  spark-ignition, 165–168, 190–191, 232
Internal Revenue Code, 28, 44
inulin, 51, 232
iodine method, 85–86
Iogen Corporation of Canada, 57, 210–214
Jerusalem artichoke, 40, 46, 48, 51–52, 83
kerosene, 4, 9, 28, 34, 35, 155, 204, 228
Kluyveromyces fragilis, 51, 60
Kluyveromyces marxianus, 51
knock sensors, 190, 197, 232
Kovarik, Bill, 8–9, 10
lactic acid, 72, 88, 98, 99, 232
lactose, 51, 59, 60, 63, 232, 239
latent heat of vaporization, 35, 103, 157, 159–160, 161, 163, 232
layout plan, distillery, 144–146
leaded gasoline, 10–11, 12
lentils, 66, 83
Liebig condensers, 139, 140–141, 142, 181
lignin, 57, 58, 63, 210, 212, 232
lime, 87, 88, 95, 99, 104, 111–112
liquefaction, 66–67, 77, 85, 92–93, 95, 232
litmus paper, 87
livestock feed, 18, 40, 49, 50, 51, 54–55, 58, 61, 97, 117, 122, 237
low wines, 106, 148–149, 233
maltase, 79, 233
maltose, 63, 72, 77, 78, 79, 85, 226, 228, 233

mannose, 69, 71, 231
mash, 101, 226, 233, 239.
See also specific crops
conventional corn mash recipe, 92–94
cookers, 113, 128–132, 228, 233
making a yeast starter, 74–76
no cooking corn mash recipe, 94
potato stock mash recipe, 94–96
mass airflow sensor (MAF), 188, 233
Material Safety Data Sheet (MSDS), 73, 88
methane, 43, 117, 118, 233
methanol, 21, 156, 161, 164, 199, 233
methyl tertiary butyl ether (MTBE), 13–14, 153, 233
Meyerhof, Otto, 69
micro-bubbler tubes, 70, 100
Microprocessor Control Units (MCU), 182, 184, 185, 188, 233. *See also* Electronic Control Units (ECU)
millet, 54, 66, 83
milling, 64–65, 92, 94, 121–122, 234, 239
milo (grain sorghum), 46, 50, 54, 55, 120
miscanthus, 57, 210
miscibility, 163–164
Mohr, J.J., 12
molasses, 46, 50–51
molecular sieves, 102, 208, 234, 239
monosaccharides (simple sugars), 45, 48, 57, 64, 69, 71, 85, 87–89, 226, 234, 237
fructose, 47, 51, 55, 60, 67, 69, 71, 79, 231, 232, 238
galactose, 63, 71
glucose, 47, 51, 55, 60, 63, 65, 66, 69, 71, 72, 79, 85, 90, 92, 93, 212, 225, 227, 228, 229, 231, 232, 233, 237, 238, 239
mannose, 69, 71, 231
sucrose, 63, 67, 79, 232, 238
Mother Earth News, 2, 4, 5–7, 8, 29, 36, 114, 125, 161, 181, 183
multi-port fuel injection, 185, 196, 234
Nader, Ralph, 182
natural gas, 13, 38, 47, 112, 118, 155, 156, 201, 204, 233
needle valves, 144, 150, 168, 177, 179, 181, 234
NextGen Biofuels Fund, 210
nitrogen oxide, 161, 162, 171, 175, 234
Novozymes, 91
nutrient addition, 99
oat, 46, 54, 55, 66
octane rating, 3, 10, 13, 14, 156, 157, 158–159, 160, 164, 189, 190, 194, 202, 234
Office of Alcohol Fuels (OAF), 161
oligosaccharides, 78, 234
oranges, 60
orchard waste, 53
Organization of Arab Petroleum Exporting Countries (OAPEC), 13
Organization of Petroleum Exporting Countries (OPEC), 3, 13
oxygen sensors, 182, 187, 200, 227, 234
packing, 53, 105, 106, 108, 134, 137–138, 143, 146, 147, 150, 234
pall-ring, 137, 138, 143, 234
saddle, 105, 106, 134, 137, 234, 237
Painter, John, 205
Painterland Farms, 39, 115, 119, 120, 205–210
pall-ring packing, 137, 138, 143, 234
papayas, 60
Passmore, Jeff, 213
peaches, 46, 48, 52, 53, 83
peanuts, 46, 83
pears, 40, 46, 48, 52, 53, 83
pectinase, 235
perforated plate columns, 30, 105, 107, 136–137, 151, 208, 235, 238
performance and fuel economy, 162–165
pH, 73, 78, 81, 92, 95, 235

starch conversion, 65, 66, 67, 68, 85
sugar extraction, 98, 99
testing for, 86–87
yeast and, 71–72
phosphoric acid, 68, 87, 88, 99, 235
pineapple, 46, 60
pitching yeast, 69, 74–76, 99–100
plantain, 60
plums, 46, 53
polysaccharides, 51, 85, 232, 235. *See also* starch
positive crankcase ventilation (PCV), 161, 193, 235
positive displacement pumps, 121, 126, 127, 130, 130–131, 165, 196–197, 228, 235
post-liquefaction, 67, 93
potatoes, 40, 46, 55, 56, 66, 83, 94–96
potentiometers, 235
power valves, 168, 171, 175, 176–177, 235
powertrain control modules (PCM), 14, 200
pre-malting, 64, 66, 235
preignition, 184, 192, 194, 232, 235
presses, 52, 96, 117, 121, 122, 237
pressure gauges, 128, 187
pressure relief valves, 125, 128, 150, 235
production need and commitment, 22–25
proof gallons, 29, 30, 35, 43, 235
proof strength, 29, 43, 89–90, 102, 106, 108, 146, 190, 198, 210, 228, 229, 234, 235–236
protease, 72, 236
puddling, 185, 236
pumpkin, 46, 60, 66, 83
pumps, 50, 70, 96, 108, 126–127, 129–132, 144, 149, 150, 152, 154, 168
accelerator, 175–176
centrifugal, 75, 95, 127, 130
diaphragm, 127, 130, 165, 228
positive displacement, 121, 126, 127, 130–131, 165, 196–197, 228, 235
slurry, 120–121
vacuum, 149, 150
quick-disconnect fittings, 126, 236
raisins, 46
raw materials. *See* feedstocks and raw materials
Reagan, Ronald, 7
reboilers, 108, 109, 118, 128, 135, 143–144, 148, 149, 152, 236
record keeping, 35
rectification, 102, 104–107, 137, 236
rectifiers, 108, 127, 135, 137, 143, 236, 237
refluxing, 107, 127, 144, 228, 233, 236
refractometers, 68, 89, 90, 99, 236
rhizomes, 58, 59, 236
rice, 40, 46, 54, 66, 83
risers, 104, 226, 236
rye, 40, 46, 54, 54, 55, 66
saccharification, 67–68, 74, 82, 85, 92, 93–94, 95, 207, 228, 236, 237
saccharometers, 67, 68, 89, 90, 99, 226, 236
Saccharomyces (cerevisiae and fragilis), 60, 71, 91, 93, 94, 212, 237
saddle packing, 105, 106, 134, 137, 234, 237
safety, 37, 87, 126, 131, 187, 193, 194, 238
Material Safety Data Sheet (MSDS), 73, 88
pressure relief valves, 125, 128, 150, 235
Sclerotinia wilt, 51
separators, 92, 207, 237
Shahbazi, Abolghasem, 58
shredders, 97, 121
Shuttleworth, John, 36
silage, 25, 50, 115, 237
simple sugars. *See* monosaccharides
slurrying, 64, 65–66, 92, 94, 95, 120–121
Small Producer Tax Credit, 44, 216, 237
Smyers, Emmerson, 6
Solar Energy Research Institute (SERI), 161

sorghum, 40, 41, 46, 57, 63, 88–89, 122, 237, 238
grain, 46, 50, 54, 55, 120
sweet, 46, 48, 50, 51, 63, 90, 96
soybeans, 46, 83, 209
space heating systems, 204
spark-ignition internal combustion engines, 165–168, 190–191, 232
spark plugs, 190–191
Special Distiller's Yeast, 91, 93, 94
specific gravity, 90, 157, 226, 232, 236, 237
spent grains, 237
spent mash, 38, 65, 72, 94, 97, 99, 108
spoilage, 41, 43, 48, 52, 60, 61, 96, 120, 215
starch, 52–55, 64, 90–100, 237. *See also specific starch crops*; starch conversion
starch conversion, 63–68
enzymes and, 76–79
fermentation, 69–70, 91–96
liquefaction, 66–67
making a yeast starter for, 74–76
milling, 64–65
post-liquefaction, 67
saccharification, 67–68, 74, 82, 85, 92, 93–94, 95, 207, 228, 236, 237
significance of yeast in, 70–74
slurrying, 64, 65–66
starch-to-sugar conversion testing, 85–86
Stargen™ enzymes, 94, 207
state and local regulations, 37–38
stillage, 49, 72, 74, 99, 116–117, 118, 127, 146, 229, 237, 238
stoichiometric ratios, 162, 187, 200, 232, 236, 237
storage, 42, 48, 53, 54, 113, 117, 118, 146, 149
alcohol, 33–34, 38, 108, 127, 139, 153–154, 164, 208
feedstocks and raw materials, 41, 54, 56, 61, 62, 63, 79, 120, 125, 144, 206
stovers, 58, 116, 237
straight vegetable oil (SVO), 21, 22
stripper columns, 108, 127, 135, 137, 141, 143, 148, 149, 237
sucrase, 79, 238
sucrose, 63, 67, 79, 232, 238
Sudan grass, 57
sugar concentration, 87–89, 98–99
sugar crops, 48–52. *See also specific crops*
sugar technique, 96–100
adding nutrients, 99
adjusting pH levels, 99
cooking, 97–98
cooling, 98
pitching yeast, 99–100
sugar concentration, 98–99
sugar extraction, 96–97
sulfuric acid, 51, 68, 87, 88, 95, 98, 99, 238
superchargers, 196–197, 238
Sustainable Agriculture Research and Education (SARE), 205
sweet potato, 40, 46, 55, 56, 83
sweet sorghum, 46, 48, 50, 51, 63, 90, 96
switchgrass, 19, 57, 58, 210
synthetic oils, 199
tanks, 120, 122–125, 147, 233. *See also* storage
fermentation, 73, 75, 94, 208, 230
mash cookers, 113, 128–132, 228, 233
tax credits, 4, 44, 200, 215–216, 237
testing procedures, 82–90
alcohol content, 89–90
feedstock, 82–85
pH, 86–87
sugar, 85–89
tetraethyl lead (TEL), 4, 8, 10–11, 182, 233, 238
thermometers, 35, 127, 128, 130, 143, 148, 149
thermostat replacement, 160, 180

throttle-body fuel injection (TBI), 166, 177, 184, 184–185, 238
Title 27 (Code of Federal Regulations), 29, 34
top dead center (TDC), 167, 189, 238
Treasury Department, 28, 31, 32, 227. *See also* Alcohol and Tobacco Tax and Trade Bureau (TTB)
true percentage of proof spirit, 217–219
tubers, 55–57, 63
  cassava, 55, 56–57
  Jerusalem artichoke, 40, 46, 48, 51–52, 83
  potato, 40, 46, 55, 56, 66, 83, 94–96
  sweet potato, 40, 46, 55, 56, 83
  yam, 46, 55, 83, 84
turbochargers, 196–197, 202, 238, 239
turbulent flow, 137, 138, 194, 238
two-stroke engines, 167–168, 198–199
United States Department of Agriculture (USDA), 61, 82, 100, 205
unleaded gasoline, 13, 14, 28, 34, 35, 104, 158, 164, 183, 194, 200, 208, 230, 231
vacuum distillation, 103, 125, 128, 149–153, 238
vacuum pumps, 149, 150
vapor pressure, 101, 103, 128, 157, 208, 238–239
venturi, 166, 169, 175, 176, 179, 183, 239
volatility, 159, 239
Volkswagen, 17, 156, 182, 184, 187, 188, 213
weir dams, 105, 136, 239
West, Daniel, 24, 53, 115
wet-milling, 239
wheat, 40, 46, 54, 55, 55, 66, 83
whey, 46, 51, 59–60, 66, 83, 239
Whiskey Rebellion (1794), 27–28
Williams, J. Craig, 205
wort, 69, 226, 233, 235, 239
  pitching yeast, 75, 76, 235
  starch conversion, 65, 66, 67, 68, 69, 93, 94, 95, 96, 97, 98, 99, 100
  testing for starch-to-sugar conversion, 74, 85
xylose, 212, 239
yam, 46, 55, 83, 84
yeast, 25, 70–76, 239
  conventional distiller's, 60, 71, 94
  estimating sugar concentration, 73–74
  ideal environment, 71–73
  Kluyveromyces (fragilis and marxianus), 51, 60
  pitching, 69, 74–76, 99–100
  Saccharomyces (fragilis and cerevisiae), 60, 71, 91, 93, 94, 212, 237
  significance of, 70–74
  Special Distiller's Yeast, 91, 93, 94
  Torula cremoris, 60
zoning, 37–38
zymase, 238, 239
zymolysis, 93

About the Author

Richard Freudenberger has been working with renewable energy and alternative transportation for over thirty years. After a stint in the New York motor press as a tech writer and automotive road tester, he joined the editorial staff of The *Mother Earth News,* where he directed the publication's Research Department, managing the Alcohol Fuel Program, coordinating the development of experimental high-mileage vehicles, and developing solar and renewable energy projects for the magazine's 624-acre Eco Village demonstration center, where he also taught renewable energy seminars. He co-founded *BackHome* magazine in 1990 and currently serves as its publisher and technical editor. He has authored or edited more than a dozen books and technical papers on green building, renewable energy, and woodworking and keeps bees on a small acreage in the mountains of Western North Carolina.

New Society Publishers

ENVIRONMENTAL BENEFITS STATEMENT

New Society Publishers has chosen to produce this book on Enviro 100, recycled paper made with **100% post consumer waste**, processed chlorine free, and old growth free.

For every 5,000 books printed, New Society saves the following resources:[1]

35	Trees
3,189	Pounds of Solid Waste
3,508	Gallons of Water
4,576	Kilowatt Hours of Electricity
5,797	Pounds of Greenhouse Gases
25	Pounds of HAPs, VOCs, and AOX Combined
9	Cubic Yards of Landfill Space

[1]Environmental benefits are calculated based on research done by the Environmental Defense Fund and other members of the Paper Task Force who study the environmental impacts of the paper industry.

NEW SOCIETY PUBLISHERS